Environmental Disruptions

NIPA® GENX ELECTRONIC RESOURCES & SOLUTIONS P. LTD.
New Delhi-110 034

About the Author

Vir Singh, Ph.D., is Professor Emeritus at the Department of Environmental Science, GB Pant University of Agriculture & Technology, India. He has more than three decades experience in teaching, research, extension, project execution, and research supervision. He has also worked as Coordinator of Liberal Education, Director of Communication, Editor-in-Chief of two monthly magazines published by the university, and Nodal Officer of Disaster Management for the Uttarakhand Government, India. He has been educated and trained in many universities and institutes in India and has also worked with many reputed organizations outside India, including International Centre for Integrated Mountain Development (ICIMOD) based in Kathmandu, Nepal, Galilee International Management Institute (GIMI) in Israel, and Friedrich-Schiller University in Germany. He has a passion for environmental science, ecology, and agroecology, and is also fascinated by eco-philosophy.

Prof. Vir Singh has published 55 books, many monographs, lab manuals, and more than 250 research papers, book chapters, and popular articles. In his book *Fertilizing the Universe* (Cambridge Scholars, 2019), he proposes a Theory of Astrobiological Evolution. His textbook on *Environmental Plant Physiology* (Taylor and Francis, CRC Press, 2020) brings to the fore a botanical strategy for a climate-smart planet. He is also a Climate Reality Leader committed to creating awareness about the ongoing climate change and its long-term implications on every walk of life. He is also actively engaged in environmental writing. His articles on vital contemporary issues are being widely published in English and Hindi dailies and e-magazines.

Environmental Disruptions

Planet Earth in the Vicious Cycle of Pollution, Global Warming, and Climate Change

Vir Singh
Emeritus Professor
Department of Environmental Science
GB Pant University of Agriculture and Technology
Pantnagar-263145, Uttarakhand, India

NIPA® GENX ELECTRONIC RESOURCES & SOLUTIONS P. LTD.
New Delhi-110 034

NIPA® GENX ELECTRONIC RESOURCES & SOLUTIONS P. LTD.

101,103, Vikas Surya Plaza, CU Block
L.S.C.Market, Pitam Pura, New Delhi-110 034
Ph : +91 11 27341616, 27341717, 27341718
E-mail: newindiapublishingagency@gmail.com
Website: www.nipabooks.com

For customer assistance, please contact
Phone: + 91-11-27 34 17 17
Fax: + 91-11-27 34 16 16
E-Mail: feedbacks@nipabooks.com

ISBN: 978-81-19002-66-5

Composed and Designed by NIPA®.

Preface

When an appropriate environment came into being, the processes of life on Earth began. The end of life would begin with the beginning of environmental disruptions. And the beginning of the environmental disruptions has already taken place and is slowly but steadily advancing toward its climax.

Once upon a time, there was an environment—the life-yielding, life-supporting, and life-enhancing environment. Now that environment is seldom encountered. Now we are unabatedly turning our life-yielding, life-supporting, and life-enhancing environment into a life-annihilating, life-discarding, and life-depressing kind of environment! In the environment, nothing is in a static state. The environment always holds itself in a state of dynamism. This dynamism is a phenomenon of life. The environment perpetuating life was not evolved in a single go. It evolved dynamically. The environment has not arrived at its gloomy state in a single go. It has arrived pollution by pollution! The environment would not die abruptly. It would die dynamically—pollution by pollution!

The Earth's environment which evolved over billions of years has been in a state of disruptions ever since the human race ushered in what is called the Industrial Age towards the middle of the 18th Century. The development of our socioeconomic systems with a paradigm shift in our lifestyles and "high" standards of living sowed the seeds of environmental disruptions.

Not only one or two components of the environment, not only a few regions of the planet, not only a few areas in a few regions but all environmental components, all regions of the Earth, all areas of all the regions of the planet are polluted. Air is polluted, water is polluted, and land and soils are polluted. Even natural sounds are polluted by noise. The life-driving and life-illuminating light is polluted by harmful radiations. Pollution is altering the structure and composition of all environmental components. Such adverse alterations are leading to unfavorable changes in the original quality of the environment, further leading to negative implications on ecosystem functions.

Environmental disruptions trigger the biosphere mechanisms leading to climate pattern disruptions. Changes in normal climate patterns are registered by the phenomenon of global warming. Global warming is one of the hottest issues our contemporary world is haplessly facing. Ever increasing accumulation of greenhouse gases in the planet's atmosphere together with other terrestrial and probable cosmic factors is paving pathways to climate disruption, which, undoubtedly, is gradually leading to assuming a state that would not allow the biosphere to maintain its ecological integrity. The ecological integrity of the biosphere is also not to break down abruptly. It would occur dynamically until and unless a state of an ecological disclimax is reached.

Environmental disruptions consistently assuming formidable proportions are all set to culminate into climate changes up to a state of climate catastrophe. If effective measures to halt environmental disruptions are not implemented at the local, national, regional, and global scale, our Living Planet would sooner or later be fixed into an irreversible state of climate change that might snatch its status of being a Living Planet.

Environmental disruptions with their causes, consequences and mitigation approaches have been collated in the book *Environmental Disruptions: Planet Earth in the Vicious Cycle of Pollution, Global Warming, and Climate Change* presents to the world a comprehensive account of the grimmest issues and most pressing challenges before the humanity. The book also invites humanity ruling over the biosphere to resolve environmental and climate-related issues and transform their environment-eclipsing activities into environmental conservation-oriented activities.

Environmental Disruptions with the flavor of a textbook is especially intended for university students, teachers, and researchers. The book can also be highly useful for all those concerned with the state of our planet and the policymakers, government departments, non-government organizations, international organizations, and voluntary groups dedicated to the cause of the environment.

My students' efforts have been pivotal for me to write books on varied environmental and ecological issues. Their boundless enthusiasm and infinite inquisitiveness in the subjects of Environmental Science I have been teaching for decades have always inspired me to embark on writing on environmental issues.

I am proud of my wife Gita, daughter Silvi, son Pravesh, son-in-law Nic, and granddaughter Avery for always creating an exhilarating and creativity-inducing environment.

I hope the book will prove useful for all those concerned with and cautious about our environment, who want to learn various aspects of the environment and climate, who teach environmental science, and who create awareness and do everything they can do to halt and reverse environmental disruptions.

Vir Singh

Contents

Abbreviations and Acronyms

µm	micrometer
$	Dollar
µmol	micromole
AFOLU	agriculture, forestry, and other land use
AIDS	acquired immunity deficiency syndrome
Ar	argon
As	arsenic
BOD	biochemical oxygen demand
Br_2	bromine
Bq	Becquerel
C	carbon
$C_{20}H_{12}$	benzo(a)pyrene
C_2F_6	hexafluoroethane
C_3F_8	octafluoropropane
$C_6H_{12}O_6$	glucose
Ca	calcium
$CaCO_3$	calcium carbonate
$CaSO_3$	calcium sulfite
$CaSO_4$	calcium sulfate
CCl_3F	trichlorofluoromethane
CCl_2F_2	dichlorodifluoromethane
$CClF_3$	chlorotrifluoromethane
CCl_2FCClF_2	1,1,2-trichloro-1,2,2-trifluoroethane (CFC-113)
$CClF_2CClF_2$	1,2-dichlorotetrafluoroethane (CFC-114)
$CClF_2CF_3$	Chloropentafluoroethane
$CHCl_2F$	Dichlorofluoromethane
$CHCl_2CF_2Cl$	1,1-difluoro-1,2,2-trichloroethane (HCFC-122)
$CHClF_2$	chlorodifluoromethane, or difluoromonochloromethane (HCFC-22)
CHF_3	trifluoromethane, or fluoroform
CH_2F_2	difluoromethane, or difluoromethylene

CH_3F	fluoromethane, or methyl fluoride, Freon-41, Halocarbon-41, HFC-41
CHF_2CF_3	pentafluoroethane
CF_4	carbon tetrafluoride, or tetrafluoromethane
Cd	cadmium
Ce	cesium
CF_4	tetrafluoromethane
CFC	chlorofluorocarbons
CH_3Hg^+	methylmercury
CH_4	methane
Ci	curie
Cl_2	chlorine
cm	centimeter
CMM	coal-mine methane
CNG	compressed natural gas
CO	carbon monoxide
CO_2	carbon dioxide
COD	chemical oxygen demand
CPCB	Central Pollution Control Board
Cr	chromium
CSIR	Council of Scientific and Industrial Research
Cu	copper
dB	decibel
DDT	dichlorodiphenyltrichloroethane
DJ	disc jockey
DNA	deoxyribonucleic acid
DO	dissolved oxygen
e^-	electrons
EPA	Environmental Protection Agency
ESP	electrostatic precipitator
etc.	etcetera
FAO	Food and Agriculture Organization
Fe	iron
FGD	flue gas desulfurization
GHG	greenhouse gas
$GtCO_2eq/yr$	gigaton carbon dioxide equivalent per year
GWP	global warming potential
Gy	grays

H_2	hydrogen
Hb	hemoglobin
HCFC	hydrofluorocarbons
H_2O	water (hydrogen monoxide)
H_2O_2	hydrogen peroxide
H_2S	hydrogen sulfide
H_2SO_4	sulfuric acid
HAB	harmful algal blooms
HC	hydrocarbon
HCl	hydrochloric acid
He	helium
HFC	hydrofluorocarbons
Hg	mercury
HNO_3	nitric acid
HYVs	high-yielding varieties
I	iodine
IARC	International Agency for Research on Cancer
ICAR	Indian Council of Agricultural Research
ICRP	International Commission on Radiological Protection
IAEA	International Atomic Energy Agency
IPCC	Intergovernmental Panel on Climate Change
K	potassium
$K_2Cr_2O_7$	potassium dichromate
kg	kilogram
L	liter
LCFS	low-carbon fuel standard
LPG	liquid petroleum gas
m	meter
Mg	magnesium
mg/L	milligram per liter
ml	milliliter
mm	millimeter
nmol	nanomole
N_2	nitrogen
N_2O	nitrous oxide
NASA	National Aeronautics and Space Administration
Ne	neon

NF_3	nitrogen trifluoride
NH_3	ammonia
Ni	nickel
NO	nitric oxide
NO_2	nitrogen dioxide
NO_3	nitrate
NOx	nitrogen oxides
N_2O	nitrous oxide (laughing gas)
NPAH	nitropolycyclic aromatic hydrocarbons
NPK	nitrogen, phosphorus, and potassium
O	nascent oxygen
O_2	oxygen
O_3	ozone
^{o}C	degrees Centigrade/ degrees Celsius
P	phosphorus
PAH	polycyclic aromatic hydrocarbon
PAN	peroxyacyl nitrate, peroxyacetyl nitrate
Pb	lead
PFC	perfluorocarbons
PM	particulate matter
pmol/mol	picomole per mole
Po	polonium
PO_4	phosphate
ppb	parts per billion
pH	power of hydrogen
ppm	parts per million
ppt	parts per trillion
Pu	plutonium
Ra	radium
rd	rads
RNA	ribonucleic acid
ROS	reactive oxygen species
SF_6	sulfur hexafluoride
SI	International System
SO_2	sulfur dioxide
SO_3	sulfur trioxide
SPM	suspended particulate matter

Sr	strontium
Sv	sieverts
Th	thorium
U	uranium
UGC	University Grants Commission
UN	United Nations
UNEP	United Nations Environment Programme
US	United States
USA	United States of America
UV	Ultraviolet
VOC	volatile organic compounds
WAIS	West Antarctic ice sheet
WHO	World Health Organization
WNV	West Nile virus
Zn	zinc
α	alpha
β	beta
γ	gamma
$Ca(OH)_2$	calcium hydroxide
$(NH_4)_2SO_4$	ammonium sulfate
$Hb(CO)_4$	carboxyhemoglobin
$Hb(O_2)_4$	oxyhemoglobin

1

Environmental Pollution

Grigoriy/ Pexels

"Environmental pollution is an incurable disease. It can only be prevented."

-Barry Commoner

We are living in a world that is in a grim state of environmental disruptions. Callous developmental activities resulting in pollution of all environmental components and gradually culminating in global warming, climate change, and species extinction are responsible for triggering the processes of environmental disruption. Pollution of air, water, and soil is rapidly assuming alarming proportions.

The very processes of socioeconomic development are now posing threat to economic progress as well as to human health. Ominous changes in global climate are posing threat to life on Earth. The irony is that the major cause of environmental disruptions is anthropogenic rather than natural. Human's materialistic progress and environmental pollution in our contemporary world go hand in hand.

Pollution: An Eclipse of the Biosphere

The Industrial Age that began in the middle of the 18th Century brought with it a 'gift' of pollution. No component of Earth's environment, no region, no area, and no ecosystem remains unaffected by environmental pollution. Our airs are polluted. Our waters are polluted. Our soils are polluted. Our foods are polluted. Everything within the stretch of our imagination is polluted. Pollution has become havoc for life on Earth. Pollution has become the most serious threat to humans and human societies. The pollution, ironically, is ruling over life!

If we get rid of the pollution or get it reduced to the extent that it ceases to be havoc to life, we need to understand the various causes, consequences, and control measures of environmental pollution. Of course, a mere understanding of various facets and control measures is not enough; we also need to implement preventive and control measures effectively at local, regional, and global levels.

Sources of Pollution

Environmental pollution, an ever-increasing 'by-product' of the Industrial Age (Fig. 1.1) has many sources to emerge from. Technological advances in almost all sectors contribute more or less to environmental pollution. Some of the sources are shown in Fig. 1.2.

Fig. 1.1: The world ushered in the Industrial Age in 1750. Humanity witnessed a paradigm shift in lifestyle with the overconsumption of natural resources and an unabated generation of waste. What could be the worst outcome of the Industrial Age is environmental pollution. The lithosphere, hydrosphere, atmosphere, and entire biosphere are all in the widening periphery of pollution. The environmental pollution is now precipitating into climate change and posing threat to the Living Planet.(Image credit: Chris LeBoutillier from Pexels)

Solid Wastes

Solid wastes mostly originate from industries, agriculture, dwelling houses, and sewage system. The wastes generated from industries include iron dust, metal pieces, leather pieces, rubber, glass fragments, etc. Agriculture creates pollutants like crop residues, animal wastes, fruit wastes, vegetable wastes, fertilizers, etc. Domestic wastes include kitchen waste, crockery, plastics, glass pieces, etc. Pollutants to get released from sewage include human and animal excreta, detergents, etc.

Liquid Wastes

Liquid wastes released from a variety of industries, residents, hospitals, and commercial establishments have an enormous contribution to environmental pollution, especially water pollution. Industrial effluents mixed with potable water often attain lethal magnitude. Liquid wastes released from chemical industries, tanneries, breweries, refineries, textile mills, paper mills, etc. contain acids, alkalies, heavy metals, and many toxic chemicals that pollute aquatic systems. Runoff from cultivated lands dissolves fertilizers, pesticides, nitrates, and heavy metals that pollute streams and groundwater sources. Liquid wastes of domestic origin include inorganic wastes (e.g. detergents, soapy water, etc.), and organic wastes (e.g. kitchen water, urine, fecal water, etc.).

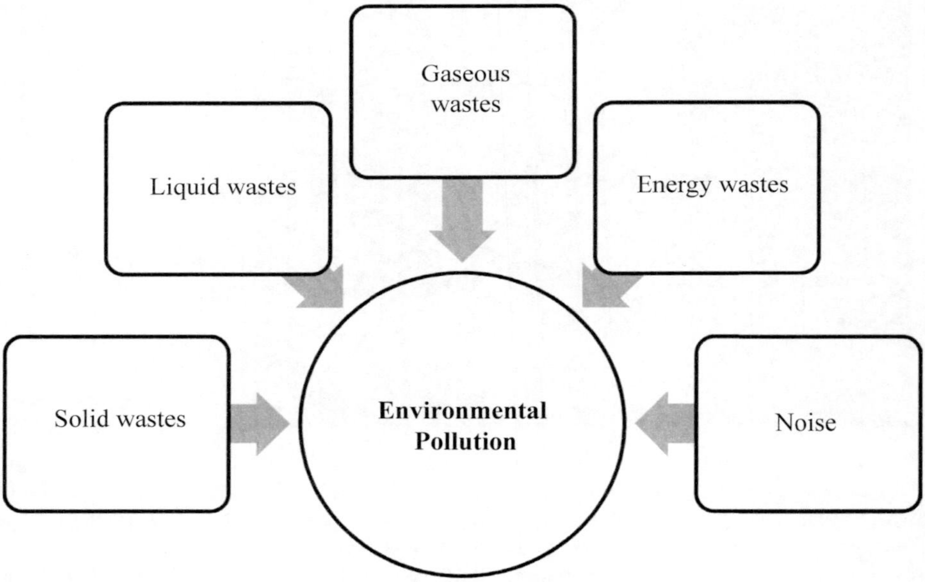

Fig. 1.2: Various sources of pollution

Gaseous Wastes

Several kinds of industries, refineries, smelters, coal-fired power plants, automobiles, and all processes requiring fossil fuels spew our deadly gaseous wastes like carbon monoxide, nitrogen oxides, sulfur dioxide, hydrogen sulfide, hydrochloric gas, etc. Mixed with air, these gases cause harmful effects on plants, animals, historical monuments, and buildings.

Energy Wastes

Among all types of polluting wastes, energy wastes are the most hazardous. These pollutants have neither mass, nor weight, and they are invisible. Heat, a 'product' of an energy source exerts heat stress on living organisms. The most hazardous one is radio-active emission. Exposure of organisms to radio-active emissions can damage their genetic makeup.

Noise

Noise or sound pollution has become an inevitable evil in technically developed societies. In our day-to-day lives, in our houses, workplaces, and public places, it is difficult to escape noise pollution. Crackers, mixers, stone crushers, radios, television, transport vehicles, public address systems, airplanes, supersonic jets, rockets, etc. create noise pollution that can rupture the eardrum and can cause irreparable mental ailments.

Various causes and effects of individual type pollution – air, water, soil, sound, and radioactive pollution – are described in the following chapters in this section.

Types of Pollutants

Let us, first of all, know: what is pollution? Pollution is an undesirable (or negative) change in the physical, chemical, and biological characteristics of the environment or a component of the environment (air, water, soil, etc.) occurring directly and/ or indirectly that may affect life adversely in one way or the other or may carry potential to harmfully affect life instantly or over some time.

Pollution is especially harmful to human beings, adversely affecting their health, natural and cultural assets, and materialistic progress, and blocking paths to a healthy, happy, and sustainable future.

Environmental pollution is owing to certain undesirable substances, chemicals, or factors called pollutants. A pollutant causes or has the potential to cause pollution. A pollutant can be a solid (e.g., dust), a liquid (e.g. a pesticide) or a gaseous (e.g., carbon monoxide) substance, or a physical factor (e.g., heat) released into the environment in such concentrations that may be or tend to be harmful to the living organisms and human beings in one way or the other. Pollutants can be divided into many categories as presented in Table 1.1.

Table 1.1: Types of Pollutants

Pollutant category	Pollutant type	Pollutant characteristics	Pollutants' examples
Origin in nature	Natural pollutants	Caused by natural agents/ happenings	Pollens, dust storms, substances of Volcanic eruption
	Anthropogenic pollutants	Caused by human beings	Chemical fertilizers, pesticides, wastes from industries and power plants, etc.
Medium	Air pollutants	Occur in the air/ atmosphere	CO, SO_2, trophospheric ozone, SPM, PAN, etc
	Water pollutants	Occur in water	Microbes, NO_3, PO_4, heavy metals, etc.
	Soil pollutants	Occur in soil	Insecticides, weedicides, heavy metals
Degradation	Biodegradable	Pollutants can be degraded or removed by means of biological/ microbial action	Domestic and municipal sewage, livestock wastes, vegetable peels, agro-wastes, etc.
	Non-biodegradable	Not degraded or removed by decomposers; undergo biomagnification	i) Wastes: aluminium cans, glass bottles, plastics; ii) Poisons: radioactive substances, heavy metals, pesticides, etc.
Occurrence in nature	Primary pollutants	Present in the same form in which they are added to environment	SO_2, NO_2, CO, hydrocarbons, pesticides, etc.
	Secondary pollutants	Formed due to reaction on primary pollutants; occur in the form different from they were released in	H_2SO_4, HNO_3, PAN, ozone, etc.
Existence in nature	Quantitative pollutants	Naturally present in environment, but also added by humans; act as pollutants if their concentration reaches a threshold value	CO_2, NO_x, etc.
	Qualitative pollutants	Not present in nature, but added due to human activity	CFCs, PAN, chemical fertilizers, pesticides, fumigants, etc.

Costs of Environmental Pollution

Environmental composition sets out conditions for survival and sustainability of life. Pollution alters environmental composition creating conditions undesirable to life. Environmental pollution costs quality of life. How much the pollution costs life would depend on the level of pollution. The effect might be invisible, temporary, small, hazardous, destructive, or catastrophic according to the level of pollution. Quoting a report of the Lancet Commission on Pollution and Health, Cohen (2017) cautions that polluted air, water, and soil kill more people than malaria, AIDS, natural disasters, wars, hunger, and smoking.

There are certain norms based on the cost of pollution that is evaluated. These are as follows:

- Medical care and treatment: Environmental pollution causes various diseases, such as asthma, jaundice, lung cancer, etc. Society has to bear the costs of medical care and treatment to respond to pollution-caused diseases.
- Loss of resource base: On account of overexploitation of the natural resources
- Erosion of biodiversity: Variety of plants, animals, and microorganisms
- Damage to agricultural resources: Foods, food crops, orchards, soil, etc.
- Corrosion of metals and metallic structures
- Damage to historical monuments and textiles
- Manpower and monetary expenditure for disposal of wastes/ pollutants and pollution control

According to the Lancet Commission on Pollution and Health, in 2015, diseases caused by air, water, and pollution, led to 9 million premature deaths, i.e., 16% of all global deaths. The global financial costs of pollution are as high as the US $ 4.6 trillion per year which is 6.2% of global economic output (Cohen 2017).

Response to Pollution

Pollution in our times is pervasive and persistent. The entire globe is plagued with pollution. No component of the environment and no part of the planet is free from pollution. Accumulation of pollutants is increasingly leading toward turning the planet uninhabitable. Effective management of environmental pollution is a pressing need and an onus on humanity.

Pollution is often avoidable and largely controllable. Unabated and unceasing human activities, such as overexploitation of natural resources, indiscriminate urbanization, overwhelming ignition of fossil fuels in industries, automobiles and power plants, etc. are steadily pushing the planet's life towards darkness. The pollution-generating processes could be restricted by imposing effective resource-conservation policies and people's participation in their implementation. However, the economic development targets coupled with fierce competition among nations for excelling in economic growth diffused the implementation programs in an undignified manner. Even the policies, programs, and protocols constituted by the UN organizations flopped before attaining their targets.

Environmental pollution can be and should be managed at three levels: local, national, and global. Some of the common measures suggested are:

- Environmental education for creating awareness among masses is a must. India's honorable Supreme Court made Environmental Studies compulsory for university students years ago. But that is not enough. Environmental education should be imparted from primary to university level and all countries should chalk out policies in this respect.
- All kinds of mass media should focus on imparting information about the state of our environment and should pass on the warning as and when necessary.
- Imposing constitutional laws should be ensured and environmental pollution must be curbed.
- Individual countries should frame effective policies and deliver their programs to heal the damage done to nature and reverse the processes that create environmental pollution.
- Specific standards for every type of pollution should be framed and violation of these standards should be strictly prohibited by laws.
- Industrial units, factories, airports, bus stands, railway stations, etc. should be established far from inhabited places.
- Nuclear testing responsible for extremely harmful radioactive pollution needs to be prohibited by international laws.
- Every country must fulfill the criteria of establishing and maintaining national parks, wildlife reserves, and biosphere reserves following the rules set up by international organizations and agencies.
- A significant proportion of the nations budget should be allocated for the development of alternative energy sources – based on inexhaustible clean energy sources – solar energy, wind energy, geothermal energy, tidal energy, etc. All nations should ensure the replacement of non-

renewable energy sources with clean energy sources in response to various UN Conventions.

- Global and regional environmental agreements have come up with several agreements and have brought about many programmers to confront environmental pollution (UNEP 2017). For example, they have fixed targets and are time-bound. These gaps might be bridged by local governing bodies. Private agencies and volunteers inculcating environmental philosophy must be encouraged to play their role.

The environment of the Earth is not just a bio-physical entity; it is a phenomenon in itself. We need to deeply understand this phenomenon, as our existence, our futures, and our sustainability are all rooted in our environment.

Summary

Pollution of air, water, and soil is rapidly assuming alarming proportions. Pollution has become the most serious threat to humans and human societies. Environmental Pollution has many sources. Technological advances in almost all sectors contribute more or less to environmental pollution. The various sources causing pollution are solid wastes, liquid wastes, gaseous wastes, energy wastes, and noise.

Pollution is an undesirable (or negative) change in the physical, chemical, and biological characteristics of the environment or a component of the environment (air, water, soil, etc.) occurring directly and/ or indirectly that may affect life adversely in one way or the other or may carry potential to harmfully affect life instantly or over some time. Environmental pollution is owing to certain undesirable substances, chemicals, living organisms (such as microorganisms), or factors called pollutants. A pollutant causes or has the potential to cause pollution. Pollution alters environmental composition creating conditions undesirable to life. It costs quality of life. The effect of pollution might be invisible, temporary, small, hazardous, destructive, or catastrophic according to the level of pollution. The costs of pollution can be determined based on medical care and treatment, loss of resource base, erosion of biodiversity, damage to agricultural resources, corrosion of metals and metallic structures, damage to historical monuments, manpower and monetary expenditure on waste disposal, pollution control, etc.

Pollution in our times is pervasive and persistent. The entire globe is plagued with pollution. No component of the environment and no part of the planet is free from pollution. Accumulation of pollutants is increasingly leading toward turning the planet uninhabitable. Effective management of environmental pollution is a pressing need and an onus on humanity. Pollution is often avoidable and largely controllable. The pollution-generating processes could be restricted by imposing effective resource-conservation policies and people's participation in their implementation.

Exercises

A. Multiple Choice Questions

1. Which of the following is a pollutant?
   a. A gaseous waste b. A liquid waste
   c. A solid waste d. All of the above
2. What is/ are the major pollution-generating agent(s) in the contemporary world?
   a. Human beings b. Ultraviolet rays
   c. Wild animals d. Greenhouse gases
3. What is not a pollutant?
   a. Heat b. Radio-active radiation
   c. Honeybees d. Ozone
4. An example of natural pollution is
   a. Volcanic eruptions b. Pollen grains
   c. Both of the above d. None of the above
5. An example of a non-biodegradable pollutant is
   a. Fruit wastes b. Plastics
   c. Vegetable wastes d. Leaf litter in a forest
6. An example of a biodegradable pollutant is
   a. Sewage b. Vegetable wastes
   c. Crop residues d. All of the above
7. Examples of anthropogenic (man-made) pollutants are
   a. Chemical fertilizers, pesticides, paper mill effluents
   b. Wastes from volcanic eruptions, pollen grains, sea sprays
   c. Particulate matter of dust storms, wastes from forest fires, UV rays
   d. None of the above

8. What category of pollutants do radioactive substances, heavy metals, and pesticides represent?
   a. Biodegradable pollutants
   b. Non-biodegradable pollutants
   c. Natural pollutants
   d. None of the above
9. The pollutants that can be degraded or removed employing biological/ microbial action are
   a. Biodegradable
   b. Non-biodegradable
   c. Man-made
   d. All of the above
10. The pollutants present in the same form in which they are added to the environment are referred to as
    a. Primary pollutants
    b. Secondary pollutants
    c. Biodegradable pollutants
    d. All of the above
11. Which of the following is a primary pollutant?
    a. CO
    b. CO_2
    c. SO_2
    d. All of the above
12. The secondary pollutants are
    a. Present in the same form in which they are added to the environment
    b. Formed due to reaction on primary pollutants; occur in the form different from they were released in
    c. Those released during a nuclear testing
    d. Relatively less harmful than the primary ones
13. The primary pollutants are
    a. Present in the same form in which they are added to the environment
    b. Formed due to reaction on primary pollutants; occur in the form different from they were released in
    c. Those released due to stratospheric ozone depletion
    d. Are always more toxic than the secondary pollutants

14. An example of a secondary pollutant is
    a. CO_2
    b. PAN
    c. CO
    d. NO_2
15. Which of the following is a mismatch?
    a. SO_2 – Primary pollutant
    b. PAN – Primary pollutant
    c. H_2SO_4 – Secondary Pollutant
    d. HNO_3 – Secondary pollutant
16. Quantitative pollutants
    a. Are naturally present in the environment, but also added by humans
    b. Act as pollutants if their concentration reaches a threshold value
    c. As at a and b above
    d. Are none of the above
17. Qualitative pollutants are
    a. Naturally present in environment, but also added by humans
    b. Are pollutants if their concentration reaches a threshold value
    c. Not present in nature, but added due to human activity
    d. Good quality pollutants
18. An example of a qualitative pollutant is
    a. CFCs
    b. Chemical fertilizers
    c. Fumigants
    d. All of the above
19. What is true about environmental pollution?
    a. It is merely anthropogenic in origin.
    b. It is often avoidable and largely controllable.
    c. It has no relationship with global warming and climate change.
    d. All the pollutants have been existing in the environment, only human actions have led to their increased concentration.
20. The headquarters of UNEP is located in
    a. Paris, France
    b. Warsaw, Poland
    c. Nairobi, Kenya
    d. Montreal, Canada

Answers: a-**d**, 2-**a**, 3-**c**, 4-**c**, 5-**b**, 6-**d**, 7-**a**, 8-**b**, 9-**a**, 10-**a**, 11-**d**, 12-**b**, 13-**a**, 14-**b**, 15-**b**, 16-**c**, 17-**c**, 18-**d**, 19-**b**, 20-**c**

B. Short-Answer Questions

1. What is pollution?
2. Define a pollutant.
3. What is the difference between primary and secondary pollutants?
4. What are the examples of natural pollution?
5. What do you mean by anthropogenic pollution?
6. What are the differences between quantitative and qualitative pollutants?
7. Categorize the following into primary and secondary pollutants: CO, CO_2, PAN, O_3, CFCs, SO_2, and NOx.
8. Differentiate between biodegradable and non-biodegradable pollutants.
9. Categorize the following examples of the pollutants into natural and anthropogenic pollutants:

   Chemical fertilizers, storm dust, pesticides, wastes from industries and power plants, pollen grains, and wastes from a volcanic eruption.
10. How much are the global financial costs of pollution, according to the Lancet Commission on Pollution and Health?

C. Long-Answer Questions

1. Describe various types of pollutants with suitable examples.
2. Discuss various sources of environmental pollution.
3. Explain the costs of pollution with examples.
4. Give an account of environmental management.
5. Write short notes on the following:
   a. Pollution as an Eclipse of the Biosphere
   b. Solid Wastes as Sources of Pollution
   c. Pollution Management

Suggested Readings

Cohen, S. 2017. The human and financial cost of pollution. *State of the Planet*, October 23, 2017.

UNEP (United Nations Environmental Program). 2017. *Towards a Pollution-Free Planet: Background Report*. Nairobi, Kenya: UNEP.

Web Resources

https://blogs.ei.columbia.edu/2017/10/23/the-human-and-financial-cost-of-pollution/

https://environmental-conscience.com/causes-effects-solutions-for-environmhttps://naturalenergyhub.com/environmental-hazards/environmental-degradation-causes-effects/ental-degradation/

https://ninequick.com/10-current-environmental-problems-in-the-world/

https://www.sciencedaily.com/releases/2019/06/190618123526.htm

https://www.unenvironment.org/beatpollution/global-response-pollution

2

Air Pollution

Frans van Heerden/Pexels

"There's so much pollution in the air now that if it weren't for our lungs there'd be no place to put it all"

– Robert Orben

Air or atmosphere constitutes one of the integral components of the environment. Among all the environmental components, the air has the highest fluidity. Therefore, the entry of a foreign substance into the air imparts a quick and severe effect on the health of the organisms. Air in the biosphere has a mixture of various gases in a dynamic proportion which is conducive to maintaining the health and sustenance of all living organisms. An alteration in the composition of the atmosphere beyond a certain limit adversely affects organisms' health, and might even be of more serious consequences for life.

Air pollution, thus, is defined as the occurrence of undesirable gases (e.g., carbon monoxide, sulfur dioxide, etc.) or particles (e.g., suspended particulate matter) in such concentrations that are deleterious to terrestrial life (vegetation, animals, and human beings) and some physical assets, like buildings, statues, and monuments. Air pollution especially affects the health of human beings living in highly modified (anthropogenic) ecosystems, e.g., cities and industrial areas.

Sources of Air Pollution

Natural sources of air pollution are not a matter of deep concern. Those have been existing ever since the living planet came into being. These sources of pollution, viz., volcanic eruptions, thunderstorms, sea sprays, pollen grains, etc., might have been essential in supporting the geographical and natural evolution of the planet. They might still be imminent for the same processes. Of course, at times natural pollution might be more dangerous than anthropogenic pollution. But, we cannot do anything to avert this 'pollution'. There are natural processes to recuperate the losses occurring due to natural pollution or to overcome the repercussions of such pollution. Moreover, natural pollution has not been increasing gradually. The natural sources of pollution have never posed a threat to life all over the globe. Their magnitude has been confined only to some defined areas of the globe.

Air pollution of anthropogenic origin, on the other hand, has been increasing for a long time. It covers the entire globe and its repercussions are extremely formidable. It is a menace for the entire globe, for all life on Earth. The human dimension of air pollution creation is much more menacing than the natural causes. Anthropogenic pollution seems to be steadily curbing the natural processes to reverse the effects of air pollution.

Stationary Sources

These sources of pollution are immobile or static and include point sources and non-point or area sources.

Point Sources: Fuel (coal and petroleum) combustion in industrial units, industrial processes, thermal power plants for electricity generation, refineries, metallurgical processes, stone crushers, solid-waste disposal at fixed locations, municipal incinerators, etc.

Non-point or Area Sources: Industrial areas, residential areas, open burning sites, forest fires, mining operations, construction sites, evaporative losses, etc.

Mobile Sources

These sources are responsible to cover larger areas to pollute through their mobility and include line sources or area sources.

Line sources: Add pollutants along narrow belts over long distances; e.g., trains, highway vehicles, channel vehicles, jet plains, etc.

Area Sources: Add smoke, dust, and gases over wide areas; e.g., bus stands, railway stations, airports, shipyards, industrial areas, cities, etc.

Industry and Thermal Power Plants

Burning of coal and petroleum in industrial units and thermal power plants leads to the production of huge amounts of pollutant gases. The pollutant gases produced are CO, CO_2, SO_2, H_2S, and hydrocarbons (HCs). Apart from these gases, enormous amounts of smoke, soot, flying ash, and other suspended particulate matter (SPM) also emit from the chimneys of various industries and factories. Metallurgical units produce clouds of dust and fumes carrying Pb, Hg, Cr, Ni, Cu, etc. A variety of chemical industries pour into the environment fumes loaded with deadly chemicals, like Cl_2, Br_2, H_2SO_4, HCl, and HNO_3 in addition to NO, H_2S, SO_2, and heavy metals, like As, Zn, Fe, Cu, Pb, etc.

Mining Operations

The mining industry needing large equipment to dig out minerals from deep layers of the Earth is very notorious in terms of adding huge amounts of dust loaded with all kinds of heavy metals. In addition, some toxic chemicals and gaseous pollutants, such as SO_2, H_2S, and NOx, might also be released during mining operations. Bulldozing and blasting operations during the implementation of mega projects such as large dam construction, and road construction in mountain areas, release huge amounts of particulate matter into the atmosphere.

Mobile Combustion Sources

Mobile combustion sources – including locomotives, automobiles, and aircraft – serve as an indistinguishable part of human lifestyles. They, however, also serve as one of the most notorious sources of atmospheric pollution, particularly in big cities. Combustion of petrol and diesel fuels in automobiles evolves various particulate lead compounds along with CO, NOx, and HCs. The largest proportion of the emitted gas when an automobile is in movement is that of CO (77.2%), followed by HCs (13.7%), and NOx (7.7%) (Pandey and Kulkarni 2006). Ozone is also produced indirectly, further adding to the pollution problem. When there is sunlight, the automobile exhausts react to form bluish or brownish fog called smog, a secondary pollutant, that hangs over cities congested with automobiles.

Agricultural Operations

The agriculture sector is one of the largest polluters of the environment. Several gases emit from crop fields applied with chemical fertilizers. These include carbon dioxide (due to increased rates of decomposition of organic matter in the soil), ammonia and nitrogen oxides (due to decomposition of nitrogenous fertilizers), and methane (due to repeated irrigation and marshy conditions, especially in rice fields). Green Revolution-type agriculture heavily relying on frequent irrigation and chemical fertilizers is especially more polluting than traditional and organic agriculture. Green Revolution agriculture depends on heavy machinery, like tractors for land preparation, combine harvesters, and threshers. These contribute to throwing particulate matter into the atmosphere.

A variety of deadly pesticides (insecticides, fungicides, herbicides, molluscicides, rodenticides, fumigants, etc.) are sprinkled in the crop fields. Having linked with nature's food chains, these deadly pesticides have extremely damaging effects on life – ranging from knocking down normal metabolic processes to inducing metabolic processes leading to cancer development. Among all the pollutants added by human beings to nature, perhaps pesticides are the ones that play a dominant role in life-annihilating processes.

In agriculture-related processes, the largest emissions occur due to deforestation and burning of forests and savannas, and the destruction of pasture lands to give way to agricultural production.

Ionizing Radiations

Any particle or electromagnetic wave carrying as much energy as will be enough to ionize or remove electrons from an atom is what is known as ionizing radiation; e.g., particle radiation from radioactive decay (Alpha- and Beta-particles and neutrons) and Gamma rays, X-rays, and far ultraviolet light. Atomic explosions, nuclear testing, and some scientific experiments are the major sources of ionizing particles and radiation.

Suspended Particulate Matter

Suspended particulate matter (SPM) comprises finely divided solid and liquid particles that disperse through the atmosphere causing pollution. Fuel combustion processes in industries, refineries, thermal power plants, automobiles, metallurgical plants, and at the domestic level are responsible for releasing SPM into the atmosphere. Mining operations, cement factories, stone crushers, agricultural processes using tractors, construction activities using bulldozers, burning of forests, etc. are other human activities adding to the SPM load in the atmosphere.

Some natural sources responsible for dispersing SPM pollutants through the air are dust storms, volcanic eruptions, sea sprays, etc. Among the natural biological contaminants in the air are pollen grains and several kinds of microorganisms (viruses, bacteria, protozoa, molds, etc.).

Indoor Pollution

Several pollutants are generated inside the home. Indoor pollution often goes unnoticed. Our homes are home to so many pollutants. Household cleaners, aerosol sprays, detergents, shoe polish, hair dyes, etc. are the pollutants readily available in all houses. Building materials (e.g., furnishings, paints, varnishes, asbestos, etc.), formaldehyde, insecticides, rodenticides, deodorants, cosmetic products, etc. might be several other causes of indoor pollution.

Millions of families in poor countries depend on firewood for cooking. This fills kitchens with smoke along with several obnoxious gases. Again, households in very large numbers depend on wood charcoal for cooking purposes and for warming the houses. Charcoal, upon burning, releases many poisonous gases, such as carbon monoxide, sulfur dioxide, nitrogen oxide, and volatile organic chemicals along with particulate matter.

House dust mites are quite common and powerful allergens inside homes. Damp walls in a house might help molds and mildews to grow that serve as a source of airborne allergenic pollutants. A damp and warm home environment invites several biological pollutants; e.g., microorganisms, mosquitoes, cockroaches, and many other types of vectors of viruses, bacteria, and protozoa.

Major Air Pollutants

The major ambient air pollutants can be divided into (i) primary air pollutants, e.g., oxides of carbon, oxides of nitrogen, oxides of sulfur, hydrocarbons, chemical fertilizers, pesticides, etc.; and (ii) secondary air pollutants, e.g., peroxy acyl nitrate (PAN), O_3, H_2SO_4, HNO_3, etc.

Primary Air Pollutants

Carbon Monoxide (CO)

Excepting carbon dioxide, CO is the most abundant gaseous pollutant in the atmosphere. Its amount in the atmosphere exceeds all other pollutant gases. This poisonous gas is produced due to the incomplete combustion of fossil fuels. In industries, factories, automobiles, kitchens, etc., where carbon fuels are used, CO is inevitably produced. Automobiles spew out CO. It is also produced inside houses if fuel-burning appliances are poorly maintained, or if there is a lack of oxygen. Cigarette smoking is also a source of CO. Nearly half of the total CO is added to the atmosphere by automobiles alone.

Marine colonial organisms (siphonophores) and some vegetation are the natural sources of CO. For example, the breakdown of photosynthetic pigments in algae releases mild amounts of CO. Forest fires are yet another source of CO. Haemoglobin breakdown among higher animals also contributes to producing CO. Bile juice is a mild source of CO.

CO is not only one of the major gaseous pollutants on Earth, the Martian and the Venus atmospheres also contain traces of CO, to the tune of 700 ppm and 17 ppm, respectively.

***Carbon dioxide* (CO_2)**

Carbon dioxide is found in the atmosphere only in traces. However, it is one of the most important gases. It is CO_2 that, through photosynthesis, is fixed into life. No CO_2, no photosynthesis. No photosynthesis, no life. Of course, chemosynthesis is an alternative to photosynthesis, most of the life on Earth is dependent on photosynthesis. CO_2, in essence, is the food of life. However, in our contemporary times, CO_2 is counted as a pollutant. Not, just another pollutant; all other pollutants contributing to the greenhouse effect and consequent climate change are measured against CO_2. Although not a poisonous gas, CO_2 is regarded as a major critical indicator of global warming and climate change. All other greenhouse gases (GHGs) have a certain value of global warming potential (GWP) in comparison to CO_2.

A major chunk of CO_2 is released as a result of fossil fuel combustion in industries, motor vehicles, thermal power plants, refineries, metallurgical plants, brick kilns, houses, etc. Deforestation, burning of forests and savannahs, agricultural operations, crop residues, decomposition of organic matter, enteric fermentation in ruminants, etc. are the other causes of CO_2 emissions. CO_2 has been building up in the Earth's atmosphere since the beginning of the Industrial Age. The current atmospheric concentration of CO_2 has crossed 420 ppm.

All the above-mentioned reasons for gradually increasing atmospheric CO_2 content are anthropogenic. It has now been proven that the climate change phenomenon is largely anthropogenic.

The release of CO_2 is a natural phenomenon of life. CO_2 (along with H_2O) is a product of respiration. But this is not an issue relating to the constantly increasing CO_2 in the atmosphere. As a result of global warming, oceans and seas are losing their CO_2 sequestration potential. Quite a large CO_2 proportion in the oceans returns to the atmosphere. This is also largely attributable to anthropogenic activities. Natural causes of CO_2 emissions, other than respiration, are earthquakes, volcanic eruptions, and organic matter decay.

Methane (CH_4)

Also known as carbon tetrahydride, hydrogen carbide, natural gas, and marsh gas, methane is found below ground as well as undersea floor and is a major constituent of natural gas. Anthropogenic activities like allowing fermentation under aerobic conditions of organic matter, such as manure, municipal solid waste, landfills, human wastes, etc. cause methane production. Rice fields in South Asia are also to blame for the generation of huge amounts of methane. Livestock management in human systems leads to the production of very large quantities of methane. Cattle, pigs, and chickens account for as much as 37% of human-induced methane production (FAO 2006).

Coal mining leads to the release of enormous amounts of methane gas known as coal-mine methane (CMM) or coalbed gas. Many countries in the world, especially Canada, the USA, and Australia are using methane gas emerging from coalmines as a source of fuel.

Among the natural sources of methane are the deep sedimentary strata and microorganisms. The breaking down of organic matter due to high temperature and pressure generates methane gas (also called thermogenic methane production) deep down in the Earth's crust. Abiotic (or inorganic) production of methane also takes place in the crystalline bedrock at deeper depths. The naturally occurring methane is produced through the biological route. Anaerobic microorganisms produce methane, a process known as methanogenesis, as follows:

$$CO_2 + 8H^+ + 8e^- \rightarrow CH_4 + 2H_2O$$

Microorganisms occurring in the rumen of ruminants, marshy lands, wetlands, and landfills involve in methanogenesis through anaerobic respiration.

Sulfur Dioxide (SO_2)

Sulfur is an element in the structure of cysteine and methionine, the two sulfur-containing acids in the composition of proteins. Sulfur, therefore, is essentially present in fossil fuels originating from plant and animal biomass. Combustion of these fossil fuels emits SO_2. Thermal power plants are responsible for a major chunk of SO_2 pollution. Sulfur is also an element (as an impurity) in metal ores. Thus smelting industries also release enormous amounts of SO_2 into the atmosphere. Fertilizer industries and sulfuric acid (H_2SO_4) manufacturing industries add significant amounts of SO_2 to the atmosphere. These sources account for as much as 75% of SO_2 emissions. Refineries and automobiles are largely responsible for the rest of the SO_2 emissions.

In the atmosphere, SO_2 oxidizes to SO_3 which, in combination with water, forms H_2SO_4 that becomes a cause of acid rain. About 60-70% of the acid rain occurs due to SO_2 and SO_3, the rest (30-40%) being due to NO_2 and NO_3.

The natural sources of SO_2 pollution include volcanic eruptions. It is believed that as high as 67% of the total SO_2 emissions are contributed by volcanoes. Forest fires are another natural source of SO_2 pollution.

Hydrogen Sulphide (**H_2S**)

Sulfur-containing fuels used in industries are the chief source of H_2S. Sewers are another anthropogenic source of H_2S. Production of H_2S by natural means is many times more than by industries. Coal pits, sulfur springs, volcanic eruptions, and vegetation and animal decaying, especially in shallow water bodies or wetlands, are the major natural sources of H_2S emissions.

Nitrogen Oxides (**NOx**)

NOx is a group of gases composed of nitrogen and oxygen. Nitrogen being an integral component of proteins in living organisms is the root cause of producing NOx upon combustion of fossil fuels which were produced by plant and animal biomass in the geological past. Nitrogen component in fossil fuels gets oxidized resulting in the formation of NOx. Thermal power plants, automobiles, nitrogen fertilizer factories, nitric acid manufacturing industry, and industrial processes, such as electroplating, welding, dynamite blasting, etc. are the major causes of NOx pollutants.

NOx also enter into the atmosphere through natural processes – by the oxidation of nitrogen and nitrogen compounds. Lightening stroke at high temperatures produces nitric oxide. This gas is also produced by bacterial oxidation of NH_3 in the soils.

Nitric oxide (NO): Combustion of fossil fuels at high temperatures leads to NO production. With a series of chemical reactions in the atmosphere, a large amount of this gas is converted into nitrogen dioxide (NO_2) which is more toxic. NO_2 combines with water vapors and forms nitric acid (HNO_3). Combined with ammonia, HNO_3 forms ammonium nitrate.

Nitrogen Dioxide (NO_2): This toxic gas is deep reddish in color and is the major constituent of photochemical smog in traffic-congested metropolitan cities. Thus, this pollutant gas is widely prevalent.

Nitrous Oxide (N_2O): Popular as a laughing gas, N_2O is a powerful greenhouse gas. This gas has also been identified as a scavenger of stratospheric ozone. Ravishankara et al. (2009) identify N_2O as 'the dominant ozone-depleting substance in the 21st century. Only about 36% of N_2O is produced thanks to anthropogenic activities, chiefly agriculture. The rest is produced through natural processes. Although the atmospheric concentration of N_2O, according to US Environmental Protection Agency, is only 0.330 ppm, as a pollutant of anthropogenic origin it matters.

According to EPA estimates as high as 29.5 million tonnes of N_2O were added to the atmosphere by the end of the first decade of the 21st Century of which 36% were due to human activity and 64% due to natural processes. Among human activities chemical fertilizer-based agriculture and livestock manure alone contribute about 42% of N_2O emissions. Runoff and leaching of fertilizers contribute about 25% of N_2O production. Biomass burning and fossil fuel combustion and other industrial processes add to 10% of N_2O emissions each. Human sewage adds to about 5% of N_2O. Nitric acid and adipic acid manufacturing industries are the major sources of N_2O emissions.

Among the natural processes leading to N_2O emissions are bacteria and fungi present in soil and marine environments. Microorganisms in the soil contribute as high as 60% N_2O, whereas in the marine environment they produce about 35% N_2O. Atmospheric chemical reactions are also responsible for the production of N_2O to the extent of about 5%.

Secondary Air Pollutants

Ozone (O_3)

In the stratosphere segment of the atmosphere, ozone serves as a shield to protect living organisms in the biosphere from harmful ultraviolet radiation. Thus, O_3 has a vital role to play in the protection of life on Earth. O_3 in the troposphere (near ground level), however, happens to be a pollutant gas. Also called 'bad ozone', O_3 in the troposphere is produced as a result of chemical reactions between NOx and volatile organic compounds (VOCs) in the atmosphere in the presence of sunlight. Emissions from fossil fuel-dependent industrial processes, refineries, thermal power plants, automobiles, gasoline vapors, etc. are the major sources of NOx and VOCs, and, therefore also the major causes of O_3 pollution.

Indoor O_3 pollution is also a matter of concern. Due to many causes, almost all families in our contemporary world face O_3 pollution in one way or the other. Outdoor O_3 concentration, air exchange rates, surface removal rates, indoor emission rates, and reactions of O_3 with chemicals in the air are some of the factors determining indoor O_3 pollution. The diurnal variations in the outdoor O_3 concentrations influence the corresponding changes in the indoor O_3 concentrations. The half-life of O_3 indoors, under normal conditions, is between 7 and 10 min and is primarily determined by air exchange and surface removal (Weschler 2001). Indoor O_3 and products of indoor O_3 are potentially dangerous for human health.

Photochemical Smog

The term "smog" is a combination of smoke and fog. However, in our environmental analysis of contemporary times, smog may not necessarily be associated with smoke and fog. It now refers to a state of atmospheric pollution often prevailing in traffic-congested cities (Fig. 2.1). The oft-used term is photochemical smog, that is, the atmospheric pollution created as a result of the reaction of light with certain pollutants in the atmosphere. Though automobile exhausts introducing nitrogen oxides into the atmosphere are the major cause creating a state of photochemical smog, coal-fired thermal plants, and other industrial units using fossil fuels also contribute to producing pollutants facilitating the emergence of photochemical smog in the presence of sunlight.

Fig. 2.1: Photochemical smog is a problem of traffic-congested cities (Image credit Nöel Puebla from Pexels)

Combustion of fossil fuels inside automobile engines (also in thermal power plants and other polluting factories) produces nitrogen oxides (NOx), particularly nitric oxide (NO) and nitrogen dioxide (NO_2). When exposed to ultraviolet radiation, NO_2 goes through a series of chemical reactions with hydrocarbons (HCs) in the atmosphere, producing a mixture of ozone, nitric acid, aldehydes, peroxy acyl nitrates (PANs), and volatile organic compounds (VOCs), and all mixed-up together, produce photochemical smog (Fig. 2.2).

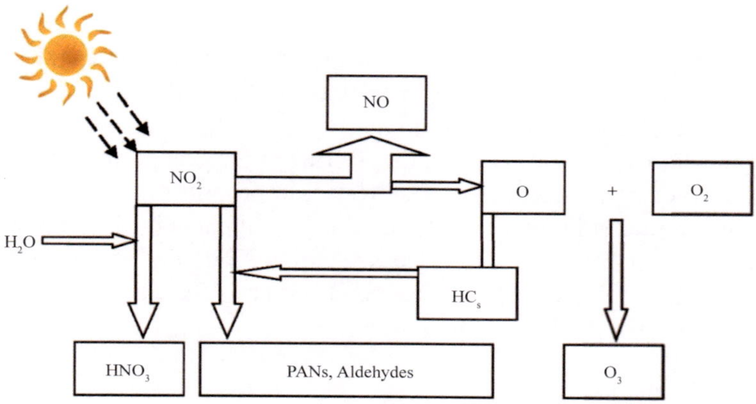

Fig. 2.2: Photochemical smog

Photochemical reactions with fossil fuel-based pollutants produce brown haze containing mainly NOx when solar radiation is intense. But when solar radiation is not intense, a grey-colored haze containing NOx, O_3, and PANs is produced.

Particulate Matter (PM)

The non-gaseous matter (solid particles and tiny liquid droplets) in the atmosphere is the particulate matter (PM). Whether the PM is settable or suspended depends on the size of the particles. Particles larger than 10 μm are settable while smaller than this remain suspended for a longer period. The suspended particulate matter (SPM) is of three types:

1. Aerosols are solid or liquid particles or both with diameters mostly less than 1 μm and behave like gases when suspended in the air;
2. Dust includes solid particles of more than 1 μm, e.g. those of coal, metals (e.g., Pb, Hg, Cu, Fe, Cr, etc.), biological agents, like pollen grains, bacterial cells, fungal spores, and dust mites; and
3. Mist includes liquid droplets more than 1 μm in diameter

The main sources of particulate matter are the industries that include mine dust, flour mill dust, iron dust, cement factories, grinding mills, etc. Jet planes emit aerosols including chlorofluorocarbons (CFCs). CFCs are also used as coolants in air conditioners and refrigerators and also as cleaning agents in the electronic industry. Thus, a substantial proportion of the SPM pollutants is anthropogenic in origin. Examples of natural SPM are fog, geyser steam, volcanic dust, forest exudates, etc.

Effects of Air Pollution on Life

Air pollution adversely affects everything, living and non-living – vegetation, animals, human beings, ecosystems (terrestrial as well as aquatic), historical monuments, and buildings – exposed to it in one way or the other. The effects could be short-term and curable, long-term, as well as permanent and incurable. Whenever and wherever it assumes toxic proportions, i.e., due to the presence of pollutants known or suspected to cause cancer, birth defects, genetic mutation, and other serious diseases among human beings when exposed to even their relatively lower concentrations in the air, it becomes extremely dangerous for public health.

Effects on Vegetation

Air pollution in general affects vegetation in various ways. The specific effect, however, depends on the species of the plants and the nature of the air pollutants.

Particulate matter (e.g., smoke, soot, dust, etc.) settle on the leaves of the plants and reduce rates of photosynthesis, affecting the overall growth of the plants.

Fluorides affect plants by inducing chlorosis, necrosis of leaf margin and tip, and leaf drop (abscission). Yellowing of leaves and slowing down of growth are the main symptoms emanating from plants exposed to fluorine pollution.

Sulfur dioxide gas is known for causing chlorosis (loss of green coloration of plant leaves) and necrosis (degeneration or death of tissues) of plants. The cell membrane is the most affected part of a living cell due to SO_2 pollution. SO_2 also causes metabolic inhibition. Lichens and mosses are especially vulnerable to SO_2 pollution.

Nitrogen oxides are the causal agents of defoliation (premature removal of leaves), lesions (pathological injury of tissues), necrosis, dieback (extensive necrosis of plant shoot from the tip downwards), and death in a variety of plants.

Ozone destroys the photosynthetic chlorenchyma tissues and thus slows down photosynthesis, resulting in slowing down plant growth. Oxidation caused by ozone produces chemical compounds that interfere with the energy system in the mitochondria of plant cells. In plant tissues, ozone dissolves in water, reacts with other chemicals, and disturbs the normal physiology of the plants. Ozone-induced compounds, according to NASA Earth Observatory, are responsible for a reduced number of flowers and fruits in plants, and also for a reduction in water use efficiency and many other physiological functions. Specific pathways of ozone-induced reactions in the plants, however, are not well established. Ozone is also known to react with lipids and proteins, causing the cell membrane to become leaky. Plants exposed to ozone pollution become more vulnerable to a variety of pests. Dicotyledons are more susceptible to ozone exposure than monocotyledons, and tobacco is the "early warning" for detecting ozone effects.

Hydrocarbons cause premature leaf fall and fruit drop (abscission). Substantial effects of hydrocarbon pollution on the Amazon forest have been detected using hyperspectral satellite images by Arellano et al. (2015). Their study reveals lower levels of chlorophyll content, increased foliar water content, leaf biochemical alterations, and leaf structural changes. Various symptoms of vegetation stresses due to hydrocarbon pollution were identified from space.

Peroxyacyl nitrates or peroxyacetyl nitrates (PANs); a family of phytotoxins, with PAN as the most abundant member of the family, in the photochemical smog, cause damage to chloroplasts and consequently impair photosynthesis, electron transport chain of respiration, enzyme activity, and anabolic processes in plants. PANs block Hill's reaction to photosynthesis. PAN-induced cell death involves reactive oxygen species (ROS) as suggested by increased intracellular H_2O_2 in PAN-exposed tobacco cells (Yukihiro et al. 2012) and the protective actions of ROS scavengers.

Effects on Animal and Human Health

Carbon monoxide: One of the toxic pollutants, carbon monoxide plays a deadly role by altering blood chemistry. It is capable to bind with the hemoglobin in the blood at a rate that is 200 times faster than the binding capacity of oxygen. The hemoglobin and carbon monoxide reaction forms carboxyhemoglobin ($Hb + 4CO \rightleftharpoons Hb(CO)_4$) as opposed to oxyhemoglobin ($Hb + 4O_2 \rightleftharpoons Hb(O_2)_4$) formed by the combination of hemoglobin and oxygen. When in bondage with carbon monoxide, the oxygen-carrying capacity of hemoglobin is substantially decreased, a situation that leads to hypoxia in body tissues. Headache, muscular weakness, exhaustion, nausea, psychomotor disturbances, etc. are health-related problems caused due to carbon monoxide pollution. If 50% of the hemoglobin in the blood is transformed into carboxyhemoglobin, it causes death due to anoxia (total depletion of oxygen).

Nitrogen oxides: The oxides of nitrogen hinder the functioning of the lungs and cause a decrease in gaseous exchange in blood. They can cause health problems such as coughing, wheezing, and bronchitis. Immunity against lung infection is also reduced due to nitrogen oxides. Lung edema, blood congestion, and eye irritation are caused in human beings when they are exposed to NO_2, which is also a mutagen.

Sulfur dioxide: It causes coughing, wheezing, phlegm, asthma, bronchitis, emphysema, etc. Children, adults with lung infections, and asthmatics are more vulnerable to SO_2 pollution. It is linked with cardiovascular problems and also carries mutagenic properties.

Chlorofluorocarbons (CFCs): Many creatures in the oceans as well as in terrestrial ecosystems suffer due to biomagnifications of CFCs. For example, CFCs' biomagnified amounts reach as high as 1400 mg/ kg of the pectoral muscles of a sea eagle. In mother's milk as high as 3.5 mg/kg of fat of CFCs have been recorded. Largely absorbed through inhalation, CFCs also enter body tissues to a lesser extent through ingestion and skin. Inhalation of CFCs in high concentration affects the central nervous system reflecting symptoms like alcohol intoxication. Headache, light-headedness, reduced coordination, convulsions, and tremors are the other health-related problems CFC pollution causes. Exposure to very high concentrations of CFC leads to disturbances in heart rhythms.

Ozone: Tropospheric ozone damages the mucous membrane even when its concentration is less than 1 ppm. It causes inflammation of breathing passage, reacts with the lung tissues, decreases the lungs' working capacity, and causes chest pain in persons exposed to its pollution.

Hydrocarbons: The burning of fossil fuels and biomass releases PAHs (polycyclic aromatic hydrocarbons) and NPAHs (nitro polycyclic aromatic hydrocarbons). Combustion of fat and tobacco also releases these aromatic hydrocarbons. PAHs and NPAHs are carcinogenic as well as mutagenic. Eye irritation, sneezing, coughing, and drowsiness are the other health problems associated with hydrocarbon pollution.

Peroxyacyl Nitrates (PAN): Human beings exposed to PAN pollution, especially in cities with high automobile and industrial emissions, suffer from eye irritation and breathing problems. PAN pollution is also linked to emphysema, lung-related problems, and impaired breathing.

Aldehydes: These chemicals cause respiratory allergies and irritation in the gastrointestinal tract. The toxic effects of aldehydes are also said to cause liver problems.

Benzo(a)pyrene: Released as a result of incomplete combustion of organic matter and tobacco smoking, benzo(a)pyrene ($C_{20}H_{12}$), a polycyclic aromatic hydrocarbon, is one of the deadly pollutants, listed by the International Agency for Research on Cancer (IARC) as a "Group 1 Carcinogen". Its diol epoxide metabolites react and bind to DNA which becomes a cause of mutations and eventually cancer.

Suspended Particulate Matter (SPM): The microscopic small solid or liquid particles suspended in the air cause irritation in the lungs and a consequent increase in lung tissue permeability. SPM aggravates the problem of chronic lung diseases and asthma. $PM_{2.5}$ are more dangerous than PM_{10} as they remain suspended in the air relatively for longer periods and are known to trigger chronic lung diseases, bronchitis and asthma. Exposure to $PM_{2.5}$ pollution can even lead to a heart attack.

Cotton dust: Cotton dust or textile fiber dust in cotton-based/ textile industries is a causal agent of pneumoconiosis, or lung fibrosis called byssinosis.

Lead: A poisonous metallic element, lead has been of great concern to human health. Breathing in the air polluted by lead causes several adverse effects on health. Lead pollution affects blood formation and basic cellular processes causing anemia. Exposure to high Pb pollution impairs brain functioning and, therefore, becomes a reason for neurological disorders, muscle paralysis, and behavioral disturbances. Lead pollution also results in increased blood pressure and hypertension. This heavy metal is also known to damage the kidney, liver, and reproductive systems. As the lead accumulates in body tissues, repeated exposure even to lower concentrations of Pb pollution has deleterious effects on human health in long run.

Silicon: Miners in silicon mines continuously get exposed to fine silicon dust particles in the air and develop a lung disease called pneumonoultramicroscopicsilicovolcanoconiosis. (This technical term makes the longest word – of 45 letters – in the English language.)

Indoor Pollution: Effects of indoor pollution aggravate with a spurt in outdoor pollution. However, indoor pollution could exclusively be because of indoor factors. When indoor sources increase the level of pollution, uninterrupted exposure to such pollution is likely to be highly deleterious. Effects on health owing to indoor air pollution are short-term, such as irritation of eyes, nasal passage, and throat, as well as long-term, such as asthma and other respiratory problems and cancer. Quite common symptoms of indoor pollution include sneezing, watery eyes, coughing, headache, fatigue, and dizziness. The severe respiratory problems occurring due to indoor pollution include inflammation of the nose (runny nose), nose bleeding (epistaxis), sore throat (pharyngitis), painful breathing (dyspnea), wheezing, severe lung infection, and pulmonary diseases. Some more severe effects of severe indoor pollution on human health are muscular pain, anorexia, nausea, vomiting, fever, rapid heartbeat, and shortness of breath.

Cigarette smoking in the residential environment invites a variety of health-related problems. Tar, acetone, nicotine, carbon monoxide, benzopyrene, and radioactive polonium-210 produced through smoking do not affect just the lungs and respiratory system of the smokers, but almost all organs and systems. Not only active smokers are affected adversely by smoking, but all family members. The nonsmoking family members turn into passive smokers when active smokers are smoking in their surroundings and tend to suffer from the same health problems as the smokers. A variety of health problems and risks due to smoking in a home environment (anywhere, of course) include anxiety, irritability, mood stimulation, dull sense of smell and taste, poor vision, smelly hair, unhealthy teeth, bronchitis, persistent coughing, loss of appetite, headache, high cholesterol, blood vessel constriction, heart diseases, increased risk of blood cancer, early menopause among women, infertility, etc.

Benzo(a)pyrene ($C_{20}H_{12}$) released with tobacco smoking is an undisputed carcinogenic compound known to cause mutation and eventually cancer. Polonium-210, a radioactive element the tobacco plants absorb from the air, is also released when one smokes tobacco. Polonium-210 emits alpha-radiation and can damage DNA and cause cancer.

Carbon monoxide is one of the worst indoor pollutants often generated by coal burning to keep houses warm during the winter season which may become a cause of death due to the reasons discussed earlier.

Effects of Air Pollution on Ecosystems

Air pollutants have long-term effects on ecosystems, both terrestrial and aquatic. There is hardly an ecosystem on Earth that is safe from the entry of air pollutants. Because of the fluidity of air, air pollutants are carried to almost all ecosystems and play their role there. Certain species in an ecosystem could be more sensitive to pollution. But even those that can withstand the direct impact of a pollutant are affected indirectly. Elimination or decreased density of vulnerable species in an ecosystem leads to alteration in ecosystem structure and composition. Biophysical changes in an ecosystem impact its functioning.

Pollutants comprising SPM deposit on canopies of forests, savannahs, and grasslands which results in the closure of stomata and decreased photosynthetic rates of the plants. This state of the environment reduces ecosystem productivity and alters the overall energy budget of terrestrial ecosystems. An increase in the ground-level ozone damages the cell membranes of the plants and, thus, inhibits the growth and development of vegetation in an ecosystem.

An increasing load of air pollutants, especially of nutrients and heavy metals, alters nutrient cycles going on in an ecosystem, which consequently impacts ecosystem composition and functions.

Once a pollutant becomes part of the producers (autotrophs), it flows through the entire food chain and food web operating in an ecosystem and affects all populations of all species of a community. Biomagnifications of toxic pollutants among the organisms along the trophic hierarchy within an ecosystem is another characteristic of an ecosystem affected by air pollution.

Entry of sulfur and nitrogen oxides into an ecosystem may cause acid deposition on vegetation and soil. These pollutants from the atmosphere also come down through acid rain resulting in soil acidification leading to the knocking down of soil microflora and fauna and impoverishment of soil fertility. The acidification of aquatic ecosystems is lethal for aquatic life. Coral reef-based aquatic ecosystems are especially vulnerable to acidification caused by sulfur and nitrogen oxides dissolved in water.

Plant nutrients in air pollution accumulate in aquatic ecosystems and trigger the processes of algal bloom, gradual extermination of aquatic animals, eventual eutrophication, and drying up of water bodies.

Aesthetic and Economic Losses

Aesthetic aspects include the quality of the environment – the aspects of the environment that are vital for happy and healthy living. Clarity, transparency, and visibility of the atmosphere have an impact on personal comfort, human moods, psychology, and creativity. Particulate matter pollution and smog rob the atmosphere of its clarity and transparency. Gaseous pollutants, like hydrogen sulfide and mercaptans, impart obnoxious odors to our environments.

Aesthetic losses and economic losses are indifferent. Diminished atmospheric clarity and reduced transparency have adverse effects on public health and economic activities, thus leading to economic losses. It is a bitter fact of our times that air pollution results in a heavy toll on the world's economies due to its negative, and often lethal, implications for public health. Millions of people the world over lose their lives due to complex, and often incurable, diseases caused by a variety of pollutants – a human agony that cannot be gauged on a monitory scale.

Economic losses as a result of air pollution are due to three major factors in the environment, viz.:

(i) soiling,

(ii) damage to ecosystems/ landscapes, and

(iii) deterioration of historical monuments and other physical assets.

Soiling means general dirtiness of the environment attributable to air pollutants, mainly particulate matter, and other pollutants that deposit on vegetation, buildings, vehicles, etc., and also due to smog. It requires frequent cleanliness of surroundings and residential environments, repeated washing of clothes, vehicles, etc. Thus, spoilage of aesthetic looks due to air pollutants also incurs economic losses.

Tropospheric ozone, peroxyacyl nitrates, sulfur dioxide, and nitrogen dioxide are the major air pollutants that destroy ecosystems and landscapes diminishing the scenic beauty of nature. Ecological damage to terrestrial ecosystems, like natural forests, savannahs, grasslands, and alpine meadows, etc. and to aquatic ecosystems, like natural lakes and coral reefs, etc. leads to the deterioration in their aesthetic appeals, keeps the tourists away from them and, at the same time, inflicts heavy losses to local, national and regional economies. The deterioration of agricultural systems due to a variety of air pollutants is another example, giving a jolt to an agro-based economy.

Air pollutants such as sulfur dioxide and nitrogen oxides have corrosive effects on buildings, including some historical monuments. Corrosion of the Statue of Liberty in the USA and of the Taj Mahal in India owing to air pollutants has been widely reported. Air pollutants damage historical monuments made of marble by transforming the marble ($CaCO_3$) into calcium sulfate and calcium nitrate. Loss to the world's many monuments is another example of economic loss attributable to air pollution.

Prevention and Control of Air Pollution

Air pollution on our globe has assumed critical proportions and has now entered into its "finale" – where it is triggering changes in global climate and posing a threat to life. Since environmental pollution is largely attributable to anthropogenic activities to fulfill the protocol of the conventional paradigm of socio-economic development, its control is also in the human hands. Pollution created due to natural causes should not be a matter of concern, as it has always been the case with the living planet. Nature has its ways and means to strike balance. We have to take the pollution created by human activities very seriously before it is too late. Natural law says that prevention is always better than cure. So, we need to take up preventive measures before we resort to controlling measures.

Preventive Measures

Selection of a Suitable Fuel

We must avoid using the fuels that pollute the atmosphere. Fuels with lower sulfur content are better as they emit lower amounts of sulfur dioxide. Further, a fuel must have a high calorific value and produce as minimum carbon emissions as possible. Natural gas, a low-carbon fuel, can replace high-carbon fuels, like diesel, petrol, kerosene oil, and coal. Liquid petroleum gas (LPG), one of the most common low-carbon fuels, should be used as a cooking fuel in the kitchen. Compressed natural gas (CNG) can be a low-carbon replacement for diesel and petrol in transport vehicles. A low-carbon fuel standard (LCFS) is a rule formulated to reduce carbon emissions caused due to overwhelming use of carbon-intensive conventional fuels using natural gas fossil fuel alternatives. These fuels are good examples to cut off carbon emissions and reduce greenhouse gases and health risks.

Suitable Site for Industrial Establishment

Industrial units and manufacturing sites must be located at a safe distance from residential areas. Applying appropriate measures, such as tall chimneys, to disperse pollutants will reduce public health risks. However, such measures would not ensure a reduction in environmental pollution but would help minimize its impact on the biotic components of the environment, including on human beings.

Alternative Energy Sources

Comparatively suitable fossil fuels (CNG, LPG, unleaded petrol, and diesel) are not a permanent solution to environmental problems. Dependence on alternative energy sources involving clean energy sources, such as solar energy, wind power, hydrogen energy, hydro-energy, etc. will be an environment-friendly and permanent solution. Energy generated from renewable and inexhaustible sources is absolutely clean with almost no ill-effect on the environment and human health. Although the total replacement of conventional energy sources with clean sources is yet to be a distant dream, more investments and a focus on research and development in this area will be highly promising for realizing a paradigm shift on the energy front.

Control Measures

We can no more afford to give air pollution a free hand. Its control is not only important but also an imperative of our times if the biosphere is to be allowed to go vibrant with ecological processes and if humanity is allowed to stay healthy, happy, and progressive. There could be three ways to control air pollution, viz.:

1. capturing the pollutants before they enter the atmosphere,
2. destruction of the pollutants, and
3. conversion of pollutants into a non-toxic or less toxic form.

Which of the above-mentioned control measures is applied depends on the nature of a pollutant and available techniques by which the pollutant could be arrested before it enters into the atmosphere, destroyed, or converted into its non-toxic or less toxic form. Some readily available techniques indicated in Fig. 2.3 are discussed under the following sub-heads:

Control of Particulate Matter

Air pollution as a result of particulate matter can be controlled either by using arresters or scrubbers.

Arresters: The arresters that remove particulate matter from the contaminated air are of many types but those commonly used are cyclonic separators and trajectory separators. PM-laden emissions with minimum moisture content from industries are separated by these methods by applying the principle of centrifugal force. These devices are especially suitable for particulate matter of coarser particle size. Extremely fine particles in particulate matter pollution are collected with the help of special filters designed for retaining PM particles of different sizes.

Electrostatic Precipitator (ESP): The most efficient filtration device is ESP which works on the principle of induced electrical charge in the PM particles flowing along with gas stream through a smokestack or chimney of an industrial unit or a fossil fuel-based power station, and collecting them on a platform charged differently, resulting in the clean gas entering into the atmosphere. Particles as small as approximately 1.0 microns, and even 0.01 microns, can be removed using ESP devices.

Scrubbers: One of the common air pollution control devices involves scrubbers. In this device, a liquid medium is used to entrap particulate matter and dissolved pollutant gases in the industrial exhaust. In coal-burning power plants, scrubbers are used to remove sulfur emissions to reduce the risks of acid rain.

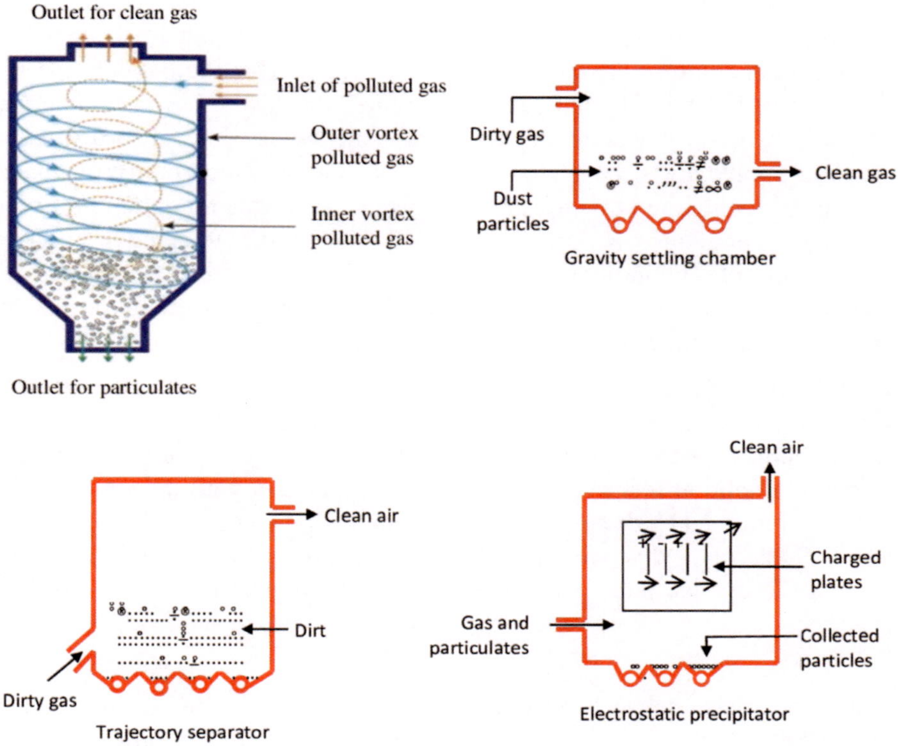

Fig. 2.3: Some devices to control air pollution

Control of Gaseous Pollutants

Pollutants in gaseous form can be controlled by applying the principles of combustion, absorption, and adsorption.

Combustion or incineration for chemically rapid oxidation of harmful gases applies to oxidizable gaseous pollutants, which can be burnt completely at high temperatures. In this process, especially VOCs and other hydrocarbon compounds in the gaseous stream are converted to CO_2 and water which are safe for health. Petrochemical, fertilizer, paint, and varnish industries especially use combustion devices to control air pollution. A special incinerator called an afterburner is used to accomplish the complete combustion of VOCs and other hydrocarbon compounds. The afterburner maintains a sufficiently high temperature and causes turbulence for proper mixing of pollutants to reduce combustion time and ensure almost complete conversion of pollutant gases into their harmless constituents.

The absorption technique involves suitable absorbent materials to adsorb certain gaseous pollutants. Transfer of a gaseous pollutant in the polluting stream is done into a contacting liquid, often water. The liquid medium serves as a solvent for the pollutant gas or it captures the gas employing a chemical reaction. Wet scrubbers or packed scrubbers are used to absorb intended gases in the exhausts. Fertilizer factories are the most common in using the absorption technique in which ammonia is captured as ammonium hydroxide in a water medium. Glass factories use this technique to remove hydrogen fluoride pollution. Some water-soluble solvents (e.g., acetone and methyl alcohol) are captured in chemical plants. Some chemical plants also apply absorption techniques to remove odor from gas streams.

Flue gas desulfurization (FGD) is an absorption process aimed at controlling sulfur dioxide emissions from fossil fuel-fired plants. The process requires wet scrubbing or dry scrubbing. Wet scrubbing involves a liquid absorbent or slurry of a solid absorbent. Upon coming into contact with the wet scrubber, SO_2 gets removed by dissolving in the liquid or slurry or by chemical reaction. Dry FGD involves dry pulverized lime or limestone. When absorption of SO_2 occurs, the dry particles are removed using baghouse filters. The dry FGD process is comparatively less costly but is appropriate only in plants using low-carbon coal.

FGD system also helps recover sulfur by converting SO_2 into useful products like H_2SO_4. Such a system is called regenerable FGD. The other system in which sulfur is not recovered is known as non-regenerable.

SO_2 in its removal process forms different products depending on the absorbent used. Regenerable FGD has an additional step of sulfur recovery.

When limestone is used: $CaCO_3 + SO_2 \rightarrow CaSO_3 + CO_2$

When lime is used: $Ca(OH)_2 + SO_2 \rightarrow CaSO_3 + H_2O$

Calcium sulfite ($CaSO_3$) reacts with water forming calcium sulfate ($CaSO_4$, gypsum). Both $CaSO_3$ and $CaSO_4$ precipitate out as slurry called FGD sludge which requires its proper disposal.

The adsorption technique uses a large porous solid surface (adsorbent) for the adsorption of selective pollutants and is suitable for handling inflammable compounds, toxic gases, and vapors using large solid surfaces. Organic and inorganic compounds with a molecular weight of more than 45 serve as good adsorbates. Activated alumina, activated carbon (heated charcoal), zeolite, silica gel, etc. are the common absorbents used in industry to control pollution. The most common absorbent is activated carbon because of its porosity and extremely high ratio of surface area to volume.

The adsorption process for the removal of targeted pollutants is physical as well as chemical. Physical adsorption involves a physical bonding force referred to as van der Waals forces between adsorbate (pollutant) molecules and adsorbent causing the former to adhere to the latter. Chemical adsorption, on the other hand, takes place by creating a chemical bond between adsorbate and adsorbent referred to as chemisorption that occurs at high temperatures. Gaseous pollutants are effectively controlled using adsorption devices in industries involving food processing, rubber processing, surface coating, dry-cleaning, degreasing, etc. Apart from getting rid of harmful gases like volatile organic compounds (VOCs) and unpleasant odors, the adsorption process is also useful in the recovery of some further usable solvents like benzene.

Control of Automobile Exhausts

Automobiles should be modified to make them more fuel-efficient. Multi-point fuel injection engines can help reduce emissions of unburnt hydrocarbons.

By using catalytic conversion filters, nitrogen oxide (NO) produced on the burning of gasoline inside the engines of vehicles can be converted into molecular nitrogen, which is safe for health.

We must always be dependent on good quality fuels to reduce a load of pollutants in the atmosphere. Unleaded petrol helps get rid of the toxic metal from automobile exhausts. The use of CNG can ensure a significant reduction in emissions from the automobile sector.

Improvement in Industrial Processes

Appropriate engineering designs can be implemented to make the instruments more fuel-efficient and industrial processes can be modified so that fuel use efficiency could be increased and greenhouse gas emissions reduced significantly. Ergonomic improvement in instruments would also increase labor efficiency and comfort in implementing industrial processes.

Summary

Air pollution is defined as the occurrence of undesirable gases (e.g., carbon monoxide, sulfur dioxide, etc.) or particles (e.g., suspended particulate matter) in such concentrations that are deleterious to terrestrial life (vegetation, animals, and human beings) and some physical assets, like buildings, statues, and monuments. Natural sources of pollution are volcanic eruptions, thunderstorms, sea sprays, pollen grains, etc. The man-made, or anthropogenic pollution sources are categorized into stationary sources (point and non-point sources), and mobile sources (line and area sources). Industry and thermal power plants, mining operations, mobile combustion sources, modern agriculture, etc. are the major sources of air pollution. Indoor pollution, often neglected, may also have grave impacts on human health. The major ambient air pollutants can be divided into primary air pollutants, e.g., oxides of carbon, nitrogen, and sulfur, hydrocarbons, chemical fertilizers, pesticides, etc.; and secondary air pollutants, e.g., peroxyacyl nitrates (PAN), O_3, H_2SO_4, HNO_3, etc. When exposed to ultraviolet radiation, NO_2, the gas produced on the combustion of fossil fuels inside an engine, goes through a series of chemical reactions with hydrocarbons (HCs) in the atmosphere, producing a mixture of ozone, nitric acid, aldehydes, peroxyacyl nitrates (PANs) and volatile organic compounds (VOCs), and all mixed-up together, produce photochemical smog.

Air pollution adversely affects everything, living and non-living – vegetation, animals, human beings, ecosystems (terrestrial as well as aquatic), historical monuments, and buildings exposed to it. Air pollution in general affects vegetation in various ways. Particulate matter settles on the leaves of the plants and reduces rates of photosynthesis, affecting the overall growth of the plants. Fluorides affect plants by inducing chlorosis, necrosis of leaf margin and tip, and leaf drop (abscission). Sulfur dioxide gas is known for causing chlorosis (loss of green coloration of plant leaves) and necrosis (degeneration or death of tissues) of plants. Nitrogen oxides are the causal agents of defoliation (premature removal of leaves), lesions (pathological injury of tissues), necrosis, dieback (extensive necrosis of plant shoot from the tip downwards), and death in a variety of plants. Ozone destroys the photosynthetic chlorenchyma tissues and thus slows down photosynthesis, resulting in slowing down plant growth. Hydrocarbons cause premature leaf fall and fruit drop (abscission). Peroxyacyl nitrates (PANs) cause damage to chloroplasts and consequently impair photosynthesis, the electron transport chain of respiration, enzyme activity, and anabolic processes in plants. Various air pollutants are hazardous to human and animal health. For example, carbon monoxide plays a deadly role by altering blood chemistry, converting hemoglobin into oxyhemoglobin. The oxides of nitrogen hinder the functioning of the lungs and cause a decrease in gaseous exchange in blood. Sulfur dioxide causes coughing, wheezing, phlegm, asthma, bronchitis, emphysema, etc. Inhalation of CFCs in high concentration affects the central nervous system reflecting symptoms like alcohol intoxication. Tropospheric ozone damages the mucous membrane even when its concentration is less than 1 ppm. Hydrocarbons PAHs (polycyclic aromatic hydrocarbons) and NPAHs (nitro polycyclic aromatic hydrocarbons) released due to fossil fuel combustion are carcinogenic as well as mutagenic. Human beings exposed to PAN pollution suffer from eye irritation and breathing problems. PAN pollution is also linked to emphysema, lung-related problems, and impaired breathing. Aldehydes cause respiratory allergies and irritation in the gastrointestinal tract. Benzo(a)pyrene released as a result of incomplete combustion of organic matter and tobacco smoking is a "Group 1 Carcinogen". SPM aggravates the problem of chronic lung diseases and asthma. Cotton dust in the cotton-based industries is a causal agent of pneumoconiosis or lung fibrosis called byssinosis. Lead pollution affects blood formation and basic cellular processes causing anemia. Exposure to high Pb pollution impairs brain functioning and, therefore, becomes a reason for neurological disorders, muscle paralysis, and behavioral disturbances.

Air pollutants have long-term effects on ecosystems, both terrestrial and aquatic. Pollutants comprising SPM deposit on canopies of forests, savannahs, and grasslands resulting in the closure of stomata and decreased photosynthetic rates. An increasing load of air pollutants, especially of nutrients and heavy metals, alters nutrient cycles going on in an ecosystem, which consequently impacts ecosystem composition and functions. Biomagnifications of toxic pollutants among the organisms along the trophic hierarchy within an ecosystem are another characteristic of an ecosystem affected by air pollution. Entry of sulfur and nitrogen oxides into an ecosystem may cause acid deposition on vegetation and soil.

Clarity, transparency, and visibility of the atmosphere have an impact on personal comfort, human moods, psychology, and creativity. Particulate matter pollution and smog rob the atmosphere of its clarity and transparency. Air pollutants such as sulfur dioxide and nitrogen oxides have corrosive effects on buildings, including some historical monuments.

Preventive measures for air pollution involve the selection of suitable fuel, a suitable site for the industrial establishment, alternative energy sources, etc. Pollution control measures require capturing of the pollutants before they enter into the atmosphere, destruction of the pollutants, and conversion of pollutants into a non-toxic or less toxic form. Air pollution as a result of particulate matter can be controlled either by using arresters (like cyclonic separators and trajectory separators) or scrubbers. The most efficient filtration device is an electrostatic precipitator (ESP) that works on the principle of induced electrical charge in the PM particles flowing along with gas stream through a smokestack or chimney of an industrial unit or a fossil fuel-based power station, and collecting them on a platform charged differently, resulting in the clean gas entering into the atmosphere. One of the common air pollution control devices involves scrubbers. In this device, a liquid medium is used to entrap particulate matter and dissolved pollutant gases in the industrial exhaust. Pollutants in gaseous form can be controlled by applying the principles of combustion, absorption, and adsorption.

Exercises

A. Multiple Choice Questions

1. Which portion of the atmosphere is particularly affected by air pollution?
   a. Troposphere
   b. Stratosphere
   c. Thermosphere
   d. All of the above on an equal basis
2. Which of the following is a stationary source of air pollution?
   a. Transport vehicles
   b. Metal smelters
   c. Both transport vehicles and metal smelters
   d. None of the above
3. Air pollution
   a. can be regarded as a local phenomenon
   b. has nothing to do with the global phenomenon
   c. involves vast areas due to the transboundary movement of air pollutants
   d. does not interact with water pollution
4. Which of the following is a correct statement?
   a. The interaction of many air pollutants can occur simultaneously
   b. The chronic form of air pollution is now a bigger threat to human life
   c. Both (a) and (b) are correct
   d. None of the above is correct
5. Who in his famous book on *The Closing Circle: Nature, Man and Technology* writes that modern technology pollutes more than the technologies it replaces?
   a. Sunderlal Bahuguna   b. Vandana Shiva
   c. Barry Commoner   d. Amartya Sen

6. The combustion of gasoline and other hydrogen fuels in automobiles, in addition to producing several primary pollutants, also produces large quantities of particulates, mainly of

   a. mercury b. arsenic
   c. copper d. lead

7. In terms of weight, nearly two-thirds of human-generated air pollution is through

   a. paper mills b. cement industries
   c. automobiles d. railway workshops

8. About 70% of CO, 44% of NOx, 33% of VOC, and 21% of particulate matter are generated into air by

   a. transportation b. tannin industries
   c. sugar factories d. poultry yards

9. The petroleum refineries, textiles, pulp and paper and industrial chemicals account for about 87% __________ and 70%__________ emissions.

   a. NO_2, SO_2 b. NO_2, H_2S
   c. NO_2, CO d. SO_2, NO_2

10. CO and SO_2 can be classified as

   a. primary and secondary air pollutants, respectively
   b. primary pollutants
   c. secondary pollutants
   d. secondary and primary pollutants, respectively

11. The pollutants for which there are legal standards or criteria for human health and for taking suitable measures for pollution control are called

   a. health-related pollutants
   b. primary and secondary pollutants
   c. standard pollutants
   d. criteria pollutants

12. Which of these pollutants are used for monitoring air quality in various Indian cities?

    a. CO, O_3, and SO_2
    b. SO_2 and NO_2
    c. SPM, RPM, SO_2, and NO_2
    d. CO_2 and SO_2

13. Of the total mass of particulate matter, dust particles account for about

    a. 10%
    b. 20%
    c. 30%
    d. 50%

14. This is a colorless gas with the characteristic smell of rotten eggs, and is emitted as a result of incomplete combustion of fossil fuels and from petroleum refining industries, natural gas plants, and petroleum units. What pollutant is this?

    a. SO_2
    b. H_2S
    c. NO_2
    d. CO

15. This is a very reactive oxidant, and because of this property, it affects the calcification of bones and teeth. What gaseous pollutant is this?

    a. Fluorine
    b. Chlorine
    c. Bromine
    d. None of the above

16. $HC + NO + O_2 \rightarrow NO_2 +$ _______

    a. PAN
    b. NO_3
    c. O_3
    d. N_2O_5

17. The first incidence of acid rain in India was recorded in 1974 in

    a. Kolkata
    b. Chennai
    c. Bengaluru
    d. Trombay and Chembur area of Mumbai

18. Radioactive fallout is on account of

    a. fuel burning
    b. ore refining
    c. waste recovering
    d. nuclear device testing

19. Nearly 50% of all CO emission originates from

    a. industries
    b. houses
    c. lakes
    d. automobiles

20. The brown air seen in traffic-congested city air is

    a. NO b. SO_2

    c. NO_2 d. NO and NO_2

Answers: 1-**a**, 2-**b**, 3-**c**, 4-**c**, 5-**c**, 6-**d**, 7-**c**, 8-**a**, 9-**d**, 10-**b**, 11-**d**, 12-**c**, 13-**d**, 14-**b**, 15-**a**, 16-**a**, 17-**d**, 18-**d**, 19-**d**, 20-**d**

B. Short-Answer Questions

1. What is a pollutant?
2 Distinguish between primary and secondary pollutants.
3. How would you define air pollution?
4. Give examples of some point sources of air pollution.
5. What is smog?
6. What is the effect of PAN on plants?
7. Mention some human diseases caused by air pollution.
8. Which pollutant gases may cause acid rain?
9. Which air pollutants can damage the ozone layer?
10. Name some devices used to control air pollution.

C. Long-Answer Questions

1. Discuss the non-point sources of air pollution.
2. How is industry a bane to the environment? Explain giving examples of the air pollutants generated by various industries.
3. What are the various ill effects of air pollution on vegetation, animals and human beings, and cultural assets? Discuss in detail.
4. Explain how can air pollution be prevented and controlled?
5. Write short notes on the following:

   (a) Indoor Pollution

   (b) Photochemical Smog

   (c) Ozone Depletion

Suggested Readings

FAO. 2006. *Livestock's Long Shadow: Environmental Issues and Options*. Rome, Italy: Food and Agriculture Organization.

IPCC. 2013. *Climate Change: The Physical Basis*. WG I, full report. 512 p.

NASA Earth Observatory. The Ozone we breath. Arellano, P., Tansey, K., Balzter, H. and Boyd, D. S. 2015. Detecting the effects of hydrocarbon pollution in the Amazon forest using hyperspectral satellite images. *Environmental Pollution*, 205 (2015): 225-239.

Pandey, B. N. and Kulkarni, G. K. 2006. *Biodiversity and Environment*. APH Publishing Corporation, New Delhi. 245pp.

Ravishankara, A. R., Daniel, J. S. and Portman, R. W. 2009. Nitrous oxide (N_2O): The dominant ozone-depleting substance in the 21st century. *Science*, 326 (5949): 123-5.

Weschler, C. J. 2001. Ozone in indoor environments: concentration and chemistry. *Indoor Air*, 10: 269-288.

Yukihiro, M., Hiramatsu, T., Bouteau, F., Kadono, T. and Kawano, T. 2012. Peroxyacetyl nitrate-induced oxidative and calcium signaling events leading to cell death in ozone-sensitive tobacco cell-line. *Plant Signalling Behaviour*, 7(1): 113-120.

Web Resources

http://scienceblog.cancerresearchuk.org/2008/08/29/radioactive-polonium-in-cigarette-smoke/
https://earthobservatory.nasa.gov/Features/OzoneWeBreathe/ozone_we_breathe3.php
https://earthobservatory.nasa.gov/Features/OzoneWeBreathe/ozone_we_breathe3.php.
https://en.wikipedia.org/wiki/Carbon_monoxide
https://mentalhealthdaily.com/2015/08/11/air-pollution-causes-brain-damage/

3

Water Pollution

Naja Bertolt/ Unsplash

"Pure water is the world's first and foremost medicine."

– A Slovakian Proverb

Ours is a Living Planet because ours is a Water Planet. The health of the Living Planet depends on the health of its water resources. Ironically, the health of its water resources is at stake. And so of its life! The waters of the Earth are in the grips of pollution. Much of the water is already polluted and the rest is increasingly becoming polluted. Water pollution, indeed, is one of the worst problems life on Earth is faced with. Water pollution is halting human progress and is phenomenally affecting all walks of human civilization. Water pollution is even putting a question mark on our sustainable future. This chapter presents the causes, magnitude, and consequences of water pollution and also suggests ways and means to prevent and control water pollution and also attempts to manage water resources.

Water is a universal solvent. It is why it is very much prone to dissolve the substances that change its composition and quality. With a change in the natural composition and quality, the water becomes unfit for rendering its functions and services the way it does when it is in pure form. When the water deviates from its original characteristics (no color, no taste, no odor, and neutral pH) and composition to a certain extent, it is said to be a state of water pollution.

What is Water Pollution?

Water pollution is defined as the mixing of some substances, be they inorganic, organic, radiological, or biological, or a factor, such as heat, which deteriorates the quality of water to the extent that it either becomes a health hazard or unfit for the intended use. When polluted, the water changes its natural physicochemical characteristics, and its role to support and nurture life alters.

Almost all surface water resources of the Earth usually contain many minerals, some organic matter, and living organisms such as bacteria, protozoa, algae, aquatic plants, fishes, etc. Water as a habitat stays healthy and water is not classified as polluted water provided it supports the lives of a variety of living organisms. A water body becomes polluted when the concentration of dissolved substances and consequently of the living organisms increases beyond a level. Then water as a habitat for a variety of living organisms deteriorates and the inhabiting organisms face adverse consequences. There are some toxic chemicals, such as pesticides, and potentially toxic heavy metals, such as mercury, arsenic, lead, etc., and harmful pathogens which cause water pollution even when in small concentrations.

Water can be occasionally polluted by natural means, such as the leaching of minerals from rocks, siltation of rivers on their paths, floods during the rainy season, mixing of plant biomass and animal wastes, etc. Such natural sources of water pollution are not a matter of deep concern. Water bodies can purify themselves. Lotic ecosystems quickly get rid of their pollution. The matter of deep concern is the pollution caused by anthropogenic activities. It is due to anthropogenic activities that water sources – not only surface water but also groundwater resources and mountain glaciers – throughout the world are now largely polluted. And water pollution is intensifying as socioeconomic progress is assuming new highs. The core of the truth is that all walks of human life today are contributing to water pollution in one way or the other. The challenge posing humanity is to prevent and control this water pollution.

Fig. 3.1: Most of the Earth is water. Most of the water is polluted.
(Image credit: Pexels by Yogendra Singh)

Sources of Water Pollution

How do the pollutants make their entry into the water? There are two major pathways. Either they flow through regular channels or enter in a scattered manner. The sources which are located at a place or are well identifiable, and drain their wastes (pollutants) into water bodies through regular channels are called point sources. Sewerage systems collecting wastes from cities, and wastewater generated in industries, like the textile industry, paper industry, sugar mills, etc. are examples of the point sources of water pollution. The scattered pollutants have no fixed location or specific channels to enter into water bodies and, thus, are known as diffuse or non-point sources. These pollutants are from non-point sources, the accumulated ones on a large area. Fertilizers and pesticides scattered throughout cultivated lands, acid rain, soil erosion in mountain areas adding turbidity to streams, etc. are examples of non-point sources.

Major sources generating pollutants introduced in water, as depicted in the figure, are discussed under the following sub-heads.

Community Wastes

Waste generation is one of the main characteristics of modern civilization. Increasing human populations and swelling urban areas are generating 'mountains' of waste which, directly or indirectly, sooner or later, are ending up as water pollutants. Industrial establishments are further adding fuel to fire. Community wastes include wastewater, human and animal excreta, food residues, cleaning agents, detergents, etc. generated in and discharged from houses, and commercial and industrial establishments. All these wastes flow through a public sewerage system. Rich in organic matter and biological pollutants, sewage is characterized by its putrescibility, which is on account of the decomposition of organic matter by microorganisms. Organic matter in the community wastes invites decomposers which contribute to depleting dissolved oxygen in the sewage.

Industrial Wastes

Industries have been mushrooming since the beginning of the Industrial Age in 1750. The entire globe is surviving under the umbrella of industries. Water resources of the planet are perhaps the worst 'victims' of the pollutants generated by industrial processes. Textile, paper, mining, food processing, metallurgical, rubber, plastic, petrochemical, and chemical industries, slaughterhouses, tanneries, distilleries, refineries, etc. discharge a large number of inorganic, radioactive, organic, and biological pollutants. Some pollutants might be extremely toxic to living beings.

Inorganic industrial pollutants include chlorides, sulfates, acids, alkalies, dyes, cyanides, and oxides of metals like mercury, chromium, iron, copper, etc. Organic industrial pollutants include carbohydrates, fats, proteins, phenols, naphtha, antibiotics, aromatic compounds, etc. A list of pollutants of great human concern being generated in different kinds of industries is presented in Table 3.1.

Table 3.1: Inorganic and organic pollutants generated by various industries

Inorganic Pollutants	Industry	Organic Pollutants
Metals, ferrous sulfate, hydrogen sulfide, ferric hydroxide, chlorides, suspended particulate matter, etc.	⟵ Mining ⟶	-
Suspended solids, iron cyanide, thiocyanate, sulfides, oxides of mercury, cadmium, copper, etc.	⟵ Iron & Steel ⟶	Oil, phenol, and naphtha
-	⟵ Food Processing ⟶	Organic matter, pathogens, etc.
Sulfides, bleaching liquors	⟵ Pulp and Paper ⟶	Cellulose, hemicellulose, lignin, tannins, etc.

Inorganic Pollutants	Industry	Organic Pollutants
-	← Pharmaceutical →	Drugs, antibiotics, organic solvents, carbohydrates, proteins and fats, etc.
Tertiary ammonium compounds, alkalis	← Soap & Detergents →	Fats, fatty acids, glycerol, polyphosphates, sulfonated hydrocarbons
Acids, alkalis, silica, phosphorus, fluorine, suspended particles	← Chemical Plants →	Aromatic compounds, organic acids, dyes, nitro compounds, etc.

Agriculture

The seventh decade of the 20th century witnessed the development of new technology for increasing food production from cultivated lands. A new revolution, popularly called Green Revolution, emerged as a guiding line for a new agriculture. New seeds of food crops, or the high-yielding varieties (HYVs) developed by plant breeders, were at the heart of the Green Revolution. The high yield potential of the HYVs could be realized only when their basic requirements of irrigation, chemical fertilizers, and pesticides were met. The Green Revolution, thus, heavily relied on chemical fertilizers and pesticides which created and are continuing to create a trail of worst-ever pollution. The Green Revolution changed the way humanity fed itself.

Chemical fertilizers and pesticides in the soil are eventually washed off through irrigation, and rainfall and mixed with the water bodies affecting aquatic life as well as human beings using water for drinking and other household purposes. Not only do chemical fertilizers and pesticides pollute surface waters, but also groundwater sources. Excessive exploitation of groundwater resources for irrigating food crops of HYVs further exacerbates the leaching of pesticides, nitrates, and heavy metals into groundwater resources.

Not only agriculture but also agriculture-based industries, such as those manufacturing chemical fertilizers and pesticides of all sorts – insecticides, fungicides, weedicides, herbicides, rodenticides, etc. – are also playing their notorious role in polluting water sources. Some of the pesticides, especially the organochlorine compounds, are persistent for a long time. They eventually enter into the food chains of the biosphere. The Green Revolution increased food production but it is also playing havoc with the health of the entire biosphere. Almost every human health problem in our contemporary world, in one way or the other, directly or indirectly, is linked to agriculture.

Thermal Pollution

Thermal pollution of water is caused by a rise in the temperature of a water body. Thermal power plants and nuclear reactors are the major sources of thermal pollution. Water is used as a coolant in thermal and nuclear power plants. In the process of cooling, the used water becomes hot. The hot water is released into the original water source which contributes to increased water temperature in a water body, often a river. A sudden temperature rise depletes dissolved oxygen and causes many undesirable organisms to perpetuate in a water body, eventually deteriorating the quality of water. An increase in water temperature, or thermal pollution, directly kills many aquatic organisms and provides appropriate conditions for many species, especially those of microorganisms and worms, to prosper in the aquatic environment. Elevated temperatures, thus, disrupt the ecological balance of aquatic habitats.

Dissolved Oxygen and Water Quality

The oxygen dissolution in water is a physical process, that is, it is present in its free molecular form, not as a compound bonded with an element. Thus, it is only the level of free molecular oxygen in the water that is referred to as dissolved oxygen (DO).

The concentration of DO in aquatic habitats is one of the basic indicators of the quality of water. DO is used as a parameter to assess the health of aquatic life in a water body. In other words, the health of the organisms inhabiting an aquatic body depends on dissolved oxygen (DO). Organisms in aquatic ecosystems need oxygen in the same way as terrestrial ones. Depletion of oxygen in water deteriorates the environment of a water body.

Water taken from resources with low oxygen concentration is supposed to be contaminated with pollutants. DO level in a water body can also indicate the type of organisms perpetuating in water. If the DO content of a water body is recorded below 8.0 mg per liter, it may be considered contaminated. If the DO level comes down to 4.0 mg per liter, water quality is considered "highly polluted."

The need for DO is not alike for all organisms. For example, shallow-water fish need much higher DO levels (up to 14 mg per liter) whereas bottom feeders, such as oysters, crabs, and worms need minimal DO amounts (1-6 mg per liter). Some creatures, such as an annelid worm *Tubifex* and some insect larvae can survive in very low DO levels. Such species are also known as indicator species to assess pollution load in the water. The importance of DO for some species can be understood with a look at Figure 3.2.

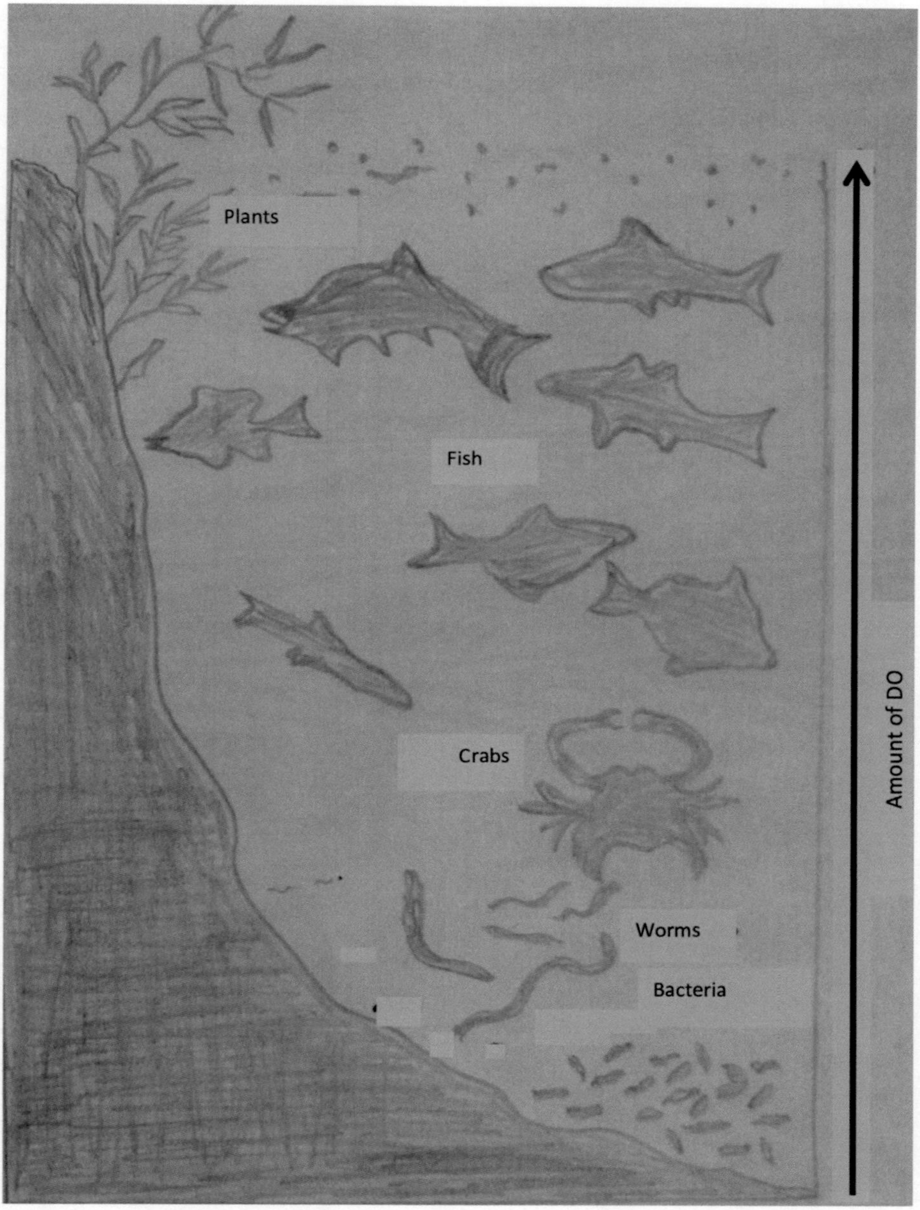

Figure 3.2: DO concentration vis-à-vis aquatic creatures

There are two sources the oxygen enters into the water from, i.e., the atmosphere and photosynthesis. Oxygen in the atmosphere continuously diffuses across water mass. Increased aeration as a result of natural factors, for example, strong winds, or due to human technical efforts increases DO diffusion rates in a water body. Photosynthesis by green plants, algae, and cyanobacteria in aquatic environments enriches water with DO amounts. DO levels due to photosynthesis increase during the daytime. The rate of CO_2 dissolution, albeit physical, is 200 times greater than that of O_2. The CO_2 available, in the presence of light, helps the photosynthesizers evolve oxygen. O_2 production and, therefore, DO levels would also depend on the availability of light which varies according to the depth of the water column. Light prevails up to about 200 m depth provided the water is clean. In turbid water, light cannot penetrate deeper layers. Photosynthesizers prevail only in the euphotic zone of a water body due to the availability of light.

DO Saturation

DO saturation means that water is holding as many O_2 molecules per unit volume as it can in equilibrium. At equilibrium, the percentage of each atmospheric gas in water will be at par with that in the atmosphere– that is, its partial pressure. In other words, water in a water body will keep on absorbing O_2 and other atmospheric gases until it attains equilibrium at complete saturation. Certain aeration-enhancing environmental factors, like wind-driven waves, accelerate O_2 absorption by water.

Deeper layers of water in a standing water body often do not get saturated with O_2 at equilibrium. This is the zone devoid of light and consequently of photosynthesis. Continuous addition of CO_2 in water due to respiration by non-photosynthetic organisms, especially animals and decomposers, prevents O_2 attain 100% saturation equilibrium. Further, wind-driving waves also do not affect deeper water zones.

The Factors affecting DO Levels

DO levels in a water body are at saturation equilibrium only under ideal environmental conditions which are very rare. Again, DO concentrations also vary from one water body to another. Temperature, pressure, and salinity are the major factors the DO levels of a water body depend upon. DO concentration decreases as the temperature of water increases. Warmer surface water would dissolve less oxygen to reach 100% air saturation than cool water in the deeper zone. DO level comes down to about zero mg per liter at 100 °C.

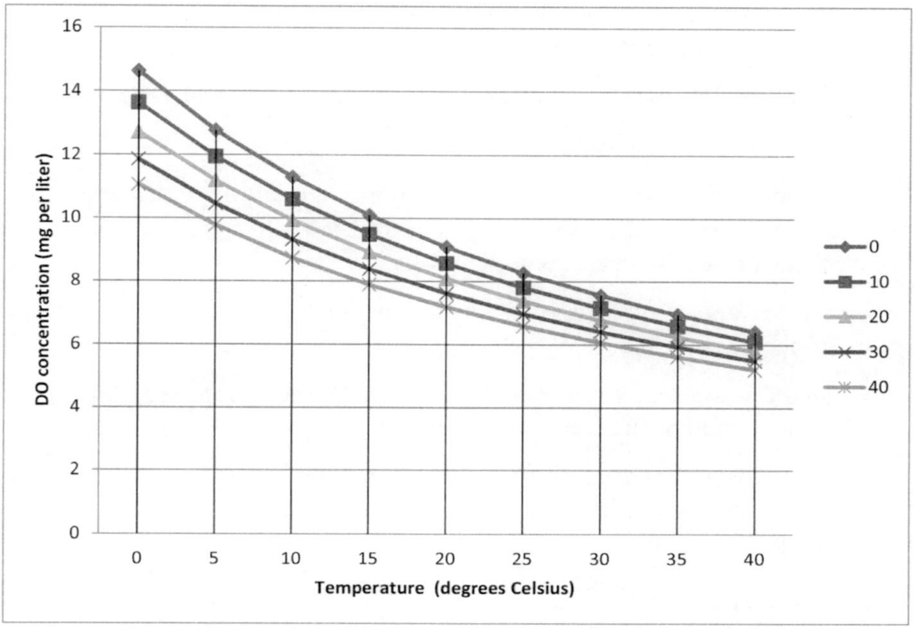

Fig. 3.3: DO concentration at varying temperatures and salinity (0, 10, 20, 30, and 40 mg/kg) levels

DO concentration decreases exponentially with an increase in water salinity. Given the same temperature and pressure, saline water will dissolve less oxygen. DO levels in a freshwater lake are higher than in marine waters at the same temperature and pressure. The DO-temperature-salinity relationship reveals that the combined effect of higher temperature and salinity results in decreased DO levels (Fig. 3.3).

An increase in pressure – atmospheric as well as hydrostatic – leads to elevated DO levels. Lower atmospheric pressure dissolves lower concentrations of oxygen. The DO concentrations, thus, are likely to be higher in water bodies at lower altitudes and plain areas than at high altitudes and in the mountains. Should the DO be at 100% air saturation at the surface, it would only be at 70%, or so, air saturation three meters below the surface.

Microorganisms, especially bacteria and fungi, also phenomenally affect DO levels. They need DO to decompose the organic matter present in water, often at the bottom of a water body. If the amount of organic matter is in excess in a standing water body (lentic ecosystem), DO concentration would decrease to a critical level.

Considering all the factors relating to DO status, and consequently health status, of the aquatic ecosystems, we can infer that:

- colder and deeper fresh waters tend to attain higher DO concentrations;
- photosynthesizers enhance DO levels;
- higher temperatures, turbidity, salinity, microbial decomposition, absence of photosynthesizers, and lack of atmospheric contact for diffusion lead to decreased DO levels;
- warmer and salty waters can also attain higher DO levels in the presence of photosynthesizers;
- Shallow waters tend to maintain higher DO levels due to atmospheric contact and constant oxygen diffusion;
- lotic ecosystems have a higher capacity to hold higher DO concentrations than lentic ecosystems due to turbulence and, consequently, enhanced aeration.

BOD and COD

DO is one of the basic determinants of water quality and the health status of aquatic life. The two other parameters determining water quality and associated with DO values are what are called biochemical oxygen demand (BOD) and chemical oxygen demand (COD).

Water pollution due to the presence of organic wastes is measured against BOD. Organic wastes in water are acted upon by decomposers which, in turn, need oxygen dissolved in water to perpetuate and accomplish decomposition.

BOD is the measure of DO needed by decomposers in the decomposition of organic matter present in water. It is expressed in mg of O_2 per liter of water. The higher the value of BOD the lower the value of DO in the water.

BOD, however, is just a measure of biodegradable organic pollutants in water. Therefore, it has some limitations. Better than BOD is COD which also takes into consideration all the oxygen-consuming substances present in water. COD is a measure of oxygen equivalent of the requirements for the complete oxidation of total organic substances, biodegradable as well as non-biodegradable, present in the water. The COD value of the polluted water, thus, is always greater than that of BOD. A comparison of various parameters of BOD and COD is made in Table 3.2.

Table 3.2: Difference between BOD and COD

Parameter	BOD	COD
Definition	Amount of DO consumed by bacteria due to decomposition of organic substances in water under aerobic conditions	Amount of DO required for the oxidation of total organic matter in water
Decomposition	A biological oxidation process	A chemical oxidation process
Test method	5210B	410.4
Test procedure	Incubation of sealed water under specific temperature sample for five days for measuring the loss of oxygen from the beginning of the test	Incubation of a closed water sample with a strong oxidant like $K_2Cr_2O_7$ in combination with boiling H_2SO_4 under specific temperature for a specified period of time
Time taken for determination	Five days	Two hours
Effect of toxic pollutants	Affected, e.g. cyanides can kill bacteria	Not affected
Effect of temperature	Decomposition process is affected	Not affected
Oxidation ability	Capable of oxidizing natural organic detritus and organic wastes	Capable to degrade industrial sewage, barring oxygen consumption of acetate
Accuracy	±10%	±2%
Permissible limits	30 mg per liter	250 to 500 mg per liter
Comparable values	< COD	> BOD

Groundwater Pollution

Groundwater is regarded as fairly a safe source of drinking water. However, this is not true. This source of water has been polluted considerably and in many areas of the world, the pollution level has assumed dangerous proportions.

Groundwater sources are polluted as a result of seepage from agricultural run-off, industrial effluents, municipal wastes, sewage channels, and landfills. In many places, groundwater contamination by nitrates and heavy metals has reached toxic levels. Overexploitation of groundwater resources for irrigation and drinking water supplies has further aggravated the pollution problem.

Marine Pollution

Pollution of oceans and seas, the largest water segment on Earth, or 97% of the Earth's waters, is a matter of deep concern for the life on Earth. If oceans and seas are polluted, it means most of the planet Earth is polluted. Pollution of rivers, streams, and other lotic ecosystems eventually ends up in marine pollution. Oceans and seas are also severely polluted by garbage and sewerage in coastal cities of the world. All sorts of chemical and nuclear wastes are also being dumped into marine ecosystems. Navigational discharges containing all kinds of pollutants, such as oils, grease, detergents, radioactive wastes, human wastes, etc. are ultimately dumped into the oceans. Capsized oil tankers, oil exploration operations, oil mining, oil refineries, etc. are notorious sources of marine pollution. Oil spills due to accidental discharges of petroleum are one of the worst kinds of pollution marine ecosystems often encounter.

Marine pollution by petroleum and petroleum products is one of the gravest global environmental issues. Mitigation of marine pollution due to the petroleum industry also appears to be almost impossible. It is because the root cause of this kind of pollution of the planet's largest ecosystem lies in the energy resource that is central to the world's socio-economic development framework: petroleum. From the exploration of oil resources to its usage, marine pollution remains the result. Marine pollution owing to petroleum is three-dimensional: incidental, accidental, and intentional.

Incidental: Extraction and isolation of crude oil from underground sources – below the Earth's surface or ocean floor – add to marine pollution. Possibilities of leakage while dealing with these initial processes of making access to crude oil sources are always there. Marine pollution also occurs incidentally due to leakage during loading, disabled tankers, drill platforms at sea, and during the transportation of petroleum through waterways. Petroleum refineries located near oceans and seas also contribute to marine pollution. Oil tankers are cleaned periodically and the washings are discharged into the oceans and seas, thus contributing to marine pollution. According to some estimates, as much as 3 million tonnes of the oil spoils marine environment thanks to this oil tanker cleaning process alone.

Accidental: The crude oil is transported from production and refinement sites to intended places throughout the world employing oil tankers or pipelines. Leakage of oil frequently occurs in both transportation modes. Accidental oil spills that have taken place so far are the worst cases of spelling out disastrous marine pollution. Some of the big oil tanker accidents that have caused havoc to marine life are as follows:

- 1967: Liberian tanker Torrey Canyon's wrecking on the Ollard Rock near England resulted in the accidental discharge of about 100,000 tonnes of oil into the sea.
- 1969: Offshore oil blew out the coast of Santa Barbara in the USA releasing about 10,000 tonnes of oil

- 1973: The oil tanker MV Cosmos Pioneer met an accident discharging about 18,000 tonnes of diesel, polluting the Gujarat coast in India.
- 1974: Transhuron, a US oil tanker, wrecked on Kilton Island in the Laccadives sea releasing some 5000 tonnes of furnace oil into the sea.
- 1976: The oil tanker Urquiola collided near the Spanish coasts of La Covuna releasing about 100,000 tonnes of petroleum.
- 1989: The oil tanker Exxon Valdez crashed in Alaska resulting in an oil spill amounting to about 11 million gallons of crude oil into Alaskan waters.
- 1992: A super oil tanker happened to meet an accident in Spain near its Northwestern coast discharging thousands of gallons of oil into the sea.

Intentional: There have been many cases when oil spills have been intentionally managed. During the Gulf war in 1991, for example, 300 million gallons of crude oil were discharged into the Persian Gulf that hit up to Saudi shores. Managed deliberately, this case of an unprecedented oil spill was a part of the war tactics enacted by Iraq against Saudi Arabia and Kuwait. Crude oil is also occasionally spilled into the seas by oil companies and oil-producing countries so that they could establish control over global oil demand and supplies.

Oceans and the seas are the ultimate "dumping grounds" of all pollutants produced by human beings. As materialistic progress is attaining new highs, marine pollution is assuming grimmer proportions. As the vastness of the oceans is being overreached by human intrusions, planetary life is sinking into deep perils of annihilation. Protecting, conserving, and ameliorating marine environments should be a priority for safeguarding biological richness, regulating global carbon, oxygen, and energy budgets, and thermoregulation of the biosphere.

Partitioning of the Spilled Oil

The spilled oil does not remain as such: it has its fate. In the process of pollution, it is partitioned in the environment in four ways:

1. Spreading
2. Evaporation
3. Solubilization
4. Residual Material

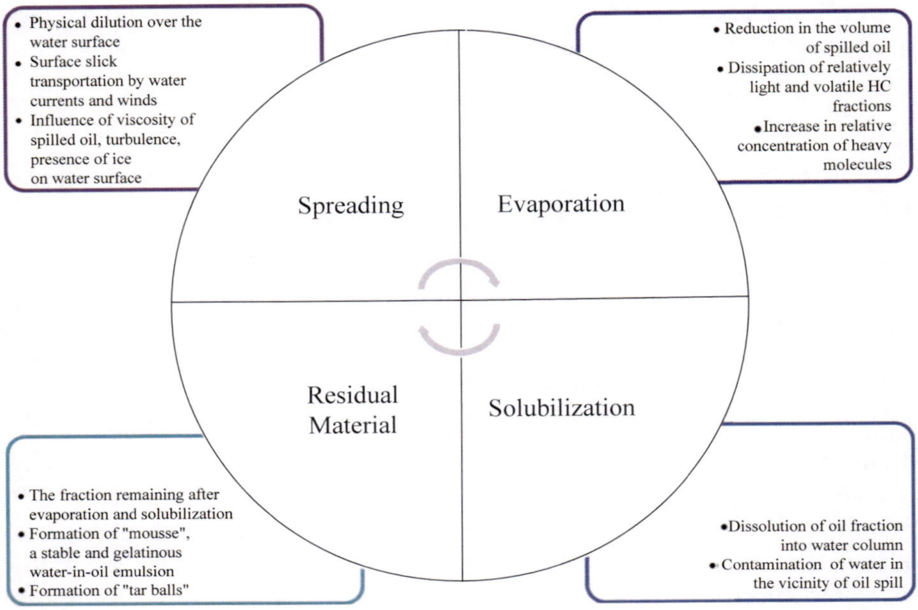

Fig. 3.4: The partitioning of the crude oil following an oil spill in marine waters

The partitioning processes of the spilled oil are depicted in Fig. 3.4. The ultimate fate of the spilled oil – after weathering by biological oxidation and photooxidation – is to be converted into "tar balls" – lumps of very dense, semisolid, asphaltic residuum (Freedman 1989). Typically C_{30} to C_{40} hydrocarbons, the balls play role in creating chronic pollution of beaches and some pelagic environments.

Effects of Water Pollution

Since water constitutes the largest proportion of the planetary environment and influences almost all life processes, its pollution imparts its effect on all the ecosystems as well as on all living organisms. The effect of water pollution largely depends on the kind of pollutants.

Effects on Aquatic Ecosystems

An aquatic ecosystem is regarded as healthy when the BOD and COD values of its water are within acceptable limits. The presence of biodegradable and non-biodegradable organic and inorganic substances brings down DO levels of the aquatic body, thus increasing the BOD and COD values of the water. When these two negative values cross a limit, aquatic organisms come under stress. As these parameters further register higher values due to the continuous addition of organic and inorganic pollutants, changes in the species composition begin to take place. Extreme limits of BOD and COD lead to the termination of sensitive aquatic species like plankton, fish, mollusks, etc. However, some tolerant species, notably annelid worms, like *Tubifex* and some insect larvae can survive amid extremely low levels of DO.

Decreased DO levels affect cold-water fish, like trout and salmon. These fish would eliminate from a coldwater system if exposed to low DO levels of 3 mg per liter for a couple of days. DO levels below 6 mg per liter are detrimental to the eggs of trout and salmon. Warm water fish namely Bluegill, White Perch, Yellow Perch, and Bass inhabit the water zone with DO above 5 mg per liter for their survival and appropriate growth. The most tolerant freshwater fish is Pike which can survive well at DO levels up to 1.5 mg per liter (Fig. 3.5).

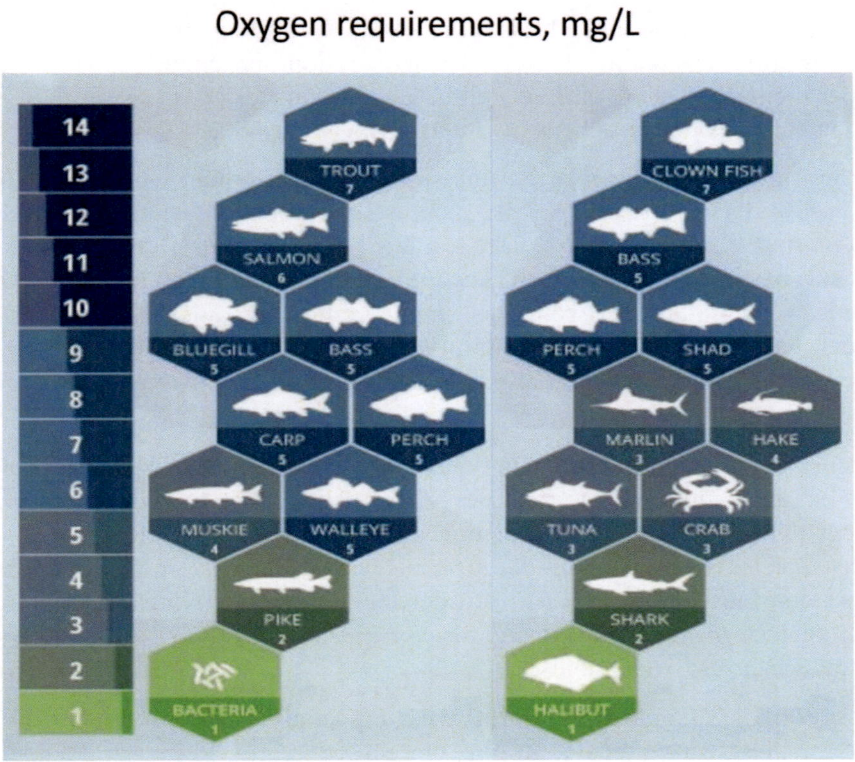

(a) (b)

Fig. 3.5: Minimum DO requirements of **(a)** fresh-water and **(b)** salt-water fish (Fondriest Environmental, Inc. "Dissolved Oxygen." Fundamentals of Environmental Measurements. 2013. Web < https://www.fondriest.com/environmental-measurements/parameters/water-quality/dissolved-oxygen/ >)

Bottom dwellers, especially microbes, face no problem with oxygen depletion. When oxygen gets completely exhausted the bacteria would use nitrates to decompose organic matter, a process known as denitrification. Should the nitrates be exhausted, the microbes would reduce sulfates. In case the organic matter accumulation is faster than its decomposition by microbes, the bottom of a water body keeps enriching in organic matter.

Saltwaters of the oceans and seas inhabit numerous organisms which also depend on DO. The capacity of saltwater to dissolve oxygen is lower than that of the freshwaters. Generally, the DO concentration of salt water is about 20% lower than that of freshwater. The organisms need desired DO levels of such waters. DO demands of different organisms might vary.

Fish species that require higher DO concentrations (often more than 5 mg per liter) to survive and do all their life activities are White Perch, Stripped Bass, and American Shad. Red Hake is quite sensitive to DO values. It escapes a zone towards the bottom of the DO concentration and comes down to 4.2 mg per liter. Many sharks thrive well in the zones of low DO concentrations. Halibut can thrive even in waters with as low a DO level as 1.0 mg per liter.

Benthic (bottom-dwelling) organisms still require minimum DO levels. The bottom-dwellers of the salt waters, such as crustaceans (lobsters and crabs), oysters, mussels, and clams require a minimum of 1-2 mg DO per liter for their survival and other activities. The minimum DO requirements of benthic organisms generally range from 1 to 4 mg/L. Benthic organisms like mussels, oysters, and clams also need a minimum of 1 to 3 mg/L of DO. It is why these organisms are found in shallower coastal waters gaining oxygen from the atmosphere and photosynthetic resources.

Not only organic and inorganic pollutants but even hot waters released from some industrial units into natural water bodies also decrease DO concentrations leading to changes in the composition of aquatic species. Certain sensitive fishery species are eliminated due to considerable decrease in DO levels in hot waters. Further, an increase in water temperatures provides appropriate conditions for microorganisms and disease-causing agents to proliferate and thus induce stress to aquatic life.

The presence of some toxic substances, especially pesticides, and heavy metals such as lead, mercury, arsenic, etc. directly eliminates several aquatic organisms in water bodies.

Biological Magnification

In a water body, a toxic or potentially toxic pollutant changes its fate if it enters into a food chain as it invariably does in an aquatic habitat. It gets accumulated in higher concentrations in the organisms of aquatic habitats and even in those of terrestrial habitats. The concentrations of the pollutant go on magnifying gradually along the trophic hierarchy of a food chain. Thus, biological magnification is the phenomenon through which certain pollutants get accumulated in organisms in increasing concentrations along the food chain. The pollutants accumulate in the tissues of the organisms.

The phenomenon of biological magnification especially applies in case of the non-degradable and persistent pollutants such as pesticides. The organochlorine compounds used as pesticides, heavy metals, and radionuclides are non-biodegradable. These are persistent for a long time. When present in an aquatic habitat, such as a pond, a lake, or a stream, such toxic pollutants inevitably make entry into plants and animals and then would be linked with the food chains prevailing in the habitat. In the tissues of an organism they are not broken down, metabolized, or excreted out. They keep on accumulating in the fatty tissues in organisms. Concentrations of the pollutants would multiply at every successive trophic level (Fig. 3.6).

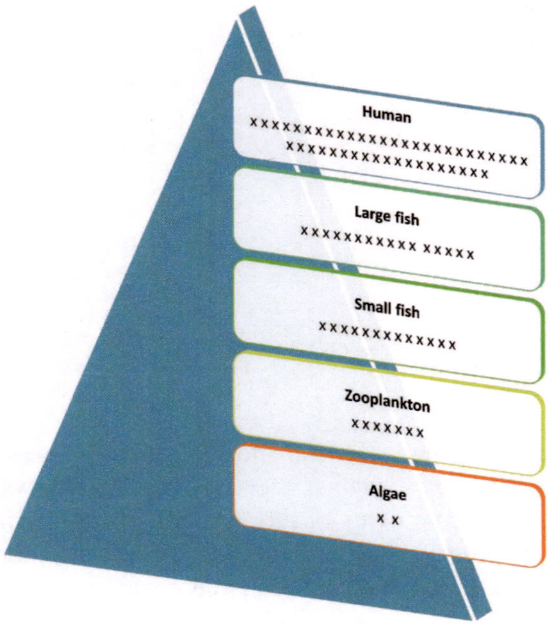

Fig. 3.6: Dramatic increase in the concentration of toxins according to successive trophic levels

Well-elaborated examples of biological magnification in the scientific literature are those of organochlorine insecticides, especially DDT (dichlorodiphenyltrichloroethane). When DDT was sprayed in residential areas to kill mosquitoes, a fraction of it made its way to a water body through runoff in the rainy season. DDT was absorbed by phytoplankton. In feeding on the phytoplankton, zooplankton also consumed DDT. From the zooplankton, DDT made entry into small fish and then into large fish. The large fish was preyed upon by the birds and the DDT entered into their tissues. Thus, DDT reached the highest stage of the food chain and its concentrations went on increasing manifold at each successive stage of the food chain. The increase in the fish-eating bird was about 5000 times more than in the phytoplankton (Fig. 3.7).

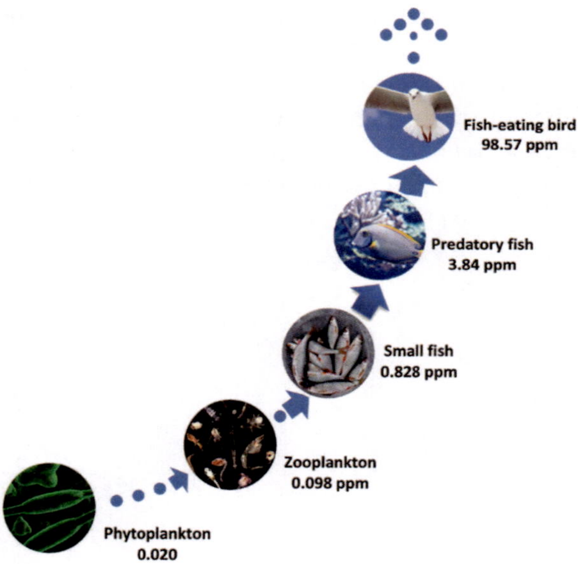

Fig. 3.7: Biological magnification of DDT in an aquatic food chain

Eutrophication

Natural water bodies are dynamic ecosystems. They are not isolated ecosystems and, therefore, cannot protect themselves against receiving nutrient flows. Nutrient enrichment of the aquatic ecosystems occurs through atmospheric flows, water run-off from adjacent areas during the rainy season, and also due to decomposition of organic matter contributed by dead and decaying plants and animals within the aquatic habitats. Nutrient enrichment of the aquatic ecosystems leads to changes in water quality as well as in biological life. The aquatic habitats, thus, change over time.

Nutrient enrichment of an aquatic ecosystem substantially enhances the growth of algae, especially blue-green algae. The impact of nitrates and phosphates on the growth of algae is especially phenomenal. Fed continuously in the nutrient-enriched environment, the growth of the algae is so profuse that it often covers the entire water body and suppresses the growth of other algae. The release of toxins and lack of oxygen due to algal bloom often kills species like fish and suppresses the growth of many other species.

The process of nutrient enrichment of a water body leading to changes in species composition, the density of populations, and loss of biodiversity is known as eutrophication. Changes in lentic ecosystems are often quicker and more peculiar than in lotic ecosystems. The latter purify themselves to a considerable extent.

Eutrophication takes place naturally over centuries as freshwater lakes age and are filled in with sediments. The submerged vegetation is an internal factor in lake ecosystem succession (Carpenter 1981). Nutrient enrichment of the water bodies is significantly influenced by human intervention. Human activities like agriculture, animal husbandry, generation of municipal and industrial wastes, and the addition of point sources and non-point sources of pollutants especially rich in nitrogen and phosphorus, to natural water bodies, have aggravated eutrophication in its intensity and extent.

Eutrophication processes begin to intensify with the profuse growth of the phytoplankton—the blue-green algae (cyanobacteria). Coverage of the water surface by algal bloom reduces water clarity and deteriorates water quality. The algal bloom limits the penetration of light due to which the growth of the plants in the littoral zone is reduced and they begin to die. Fed by essential nutrients, the algal bloom increases its photosynthetic rates that cause depletion of dissolved CO_2 increasing the pH of water during the daytime to an extreme level. An increase in the pH of the water can 'blind' organisms relying on the perception of dissolved chemical cues for their survival by impairing their chemosensory abilities (Turner and Chislock 2010). Upon the death of the algal bloom, microbial decomposition occurs which is followed by severe DO depletion and consequent hypoxic and anoxic conditions.

There are yet some algal blooms, often referred to as harmful algal blooms (HABs) that release toxins, e.g. microcystin and anatoxin-a, that spoil water quality posing risks to public health and are detrimental to fisheries. Associated with HABs in freshwaters are the cyanobacteria among which some like *Anabaena, Oscillatoria, Microcystis, and Cylindrospermopsis*, tend to dominate when the freshwater ecosystems are rich enough in nutrients, N-P ratios are low, temperatures are high and light levels are low. Serious toxic effects of the algal bloom of cyanobacteria on wildlife, livestock, and human beings have also been observed (Chislock et al. 2013).

Freshwater resources are increasingly becoming scarce especially due to burgeoning human populations, anthropogenic activities, and ongoing climate change. Eutrophication is further aggravating the problems and posing threat to all walks of life. Protection of fast-diminishing water resources from eutrophication and all processes of pollution is one of the most pressing environmental issues of our times. Control and management of cultural eutrophication employing effective and workable policies and laws and programs need to be implemented at all levels: local, national, regional, and global.

Effects on Human Health

Human health is largely subject to the quality of water. Quality of water not only matters for drinking purposes but also for bathing, all types of domestic chores, domestic and wild animals, and irrigation purposes. Water quality, in essence, is an issue in all human systems. Human health is highly susceptible to water pollution. Most of the diseases humanity suffers from the world over are attributable to the use of polluted water.

Water polluted due to sewage is sure to contain pathogens like viruses, bacteria, protozoa, and worms. The use of such contaminated water causes what we often refer to as water-borne diseases, such as amoebiasis, diarrhea, jaundice, typhoid, cholera, etc.

Serious health problems are surely caused when water contaminated with heavy metals is used. Water contaminated by cadmium causes Itai Itai disease, also called ouch-ouch disease, a painful disease of bones and joints. Cadmium consumption beyond a limit, as it may happen due to drinking cadmium-contaminated water, may cause liver and lung cancer. Bluish lines around the gums, loss of muscle power, anemia, and headache are the other human health problems cadmium-polluted water can cause.

The oft-quoted disease caused by the consumption of fish from a water source contaminated with mercury is the Minamata disease, a neurological syndrome, reported in Japan in 1956. The disease was named after Minamata bay on the west coast of Kyushu island in Japan which was contaminated with mercury used in a polythene factory located near the bay. A mercury compound was used in the reaction to produce vinyl chloride, a common plastic material. The leftover mercury and other wastes from the polythene factory were dumped into the bay. The mercury settled in the sediment at the bottom of the bay was reacted upon by microorganisms which converted it into a more toxic organic form – methylmercury, CH_3Hg^+. This poisonous compound eventually entered the aquatic food chain. Among fish of the bay, the methyl mercury underwent biomagnification. When the contaminated fish were eaten by people, the bio-magnified mercury poisoned them. Many people died and many of them developed the symptoms of mercury poisoning: ataxia, numbness in hands and feet, general muscle weakness, damage to speech and hearing, etc. The mothers who ate mercury-contaminated fish gave birth to children with symptoms of mercury poisoning.

Groundwater pollution is caused mainly due to seepage from industrial effluents and run-off from agricultural lands. Water polluted with nitrates is especially fatal for infants. Upon reaction with hemoglobin in the blood, nitrates form methemoglobin which does not transport oxygen to body tissues. This health problem is called methemoglobinemia. In infants, the insufficient release of oxygen due to excess methemoglobin causes hypoxemia and skin color turns bluish. This condition, therefore, is called a blue baby syndrome.

Overexploitation of groundwater resources may result in the entry of arsenic from soil and rock sources into water and poison it. The use of arsenic-contaminated water leads to what is known as black foot disease. In addition, arsenic is also known to cause lung and skin cancer, peripheral neuritis, hyperkeratosis, and diarrhea. In India, arsenic-polluted water was first of all noticed in 1983 in the 24-Pargana district of West Bengal. Later on, it was found affecting areas of the state of Bihar. The problem of arsenic-polluted water is rampant in Bangladesh. Thailand, Vietnam, Mexico, China, Argentina, and Chile have also been reported facing problems arising out of arsenic pollution to some extent.

Overexploitation of groundwater also leads to water pollution by excessive fluorides. Digging deeper borewells pumps out water from the aquifers containing high fluoride concentrations. Fluoride excess in potable water causes what is called skeletal fluorosis characterized by teeth deformity, stiff and painful joints, and hardening of bones. Non-skeletal fluorosis results in gastrointestinal and neurological problems. India and China are the main sufferers of fluoride problems. This problem is endemic to 19 Indian states and at least 25 countries of the world are facing the problem arising out of drinking fluoride-contaminated water affecting millions of people.

Depolluting the Water

Water needs to sustain itself in the form in which it continues to support life, sustain life, and enhance life. Water needs to be managed in such a way that it keeps away from getting polluted. Where water needs to be used in human-regulating systems, it needs to be made free from pollution through its treatments before it is reused for economic activities like irrigation and/ or poured back into streams, rivers, lakes, or oceans.

Sewage Treatment

Sewage is one of the prime management systems of urban life the world over. Its treatment is a must for maintaining the reuse of water and for saving natural water sources from getting polluted. This can be treated in three steps (Fig. 3.8).

Primary Treatment Stage

Suspended particles and large-sized objects present in the sewage are removed by simple physical methods, like filtration and sedimentation.

Secondary Treatment Stage

Oxygen is supplied to fulfill the needs of the bacteria responsible for the decomposition of organic matter the sewage is rich in. Shallow stabilization or oxidation tanks are used to carry out the decomposition process, i.e., biological oxidation. The process results in the biosolid material – sludge – and CO_2 is released. The sludge is repeatedly aerated to promote its oxidation. To ensure a continuous supply of oxygen, green algae are grown in the upper zone that receives light. Aeration takes place due to the release of oxygen during photosynthesis the algae involve in.

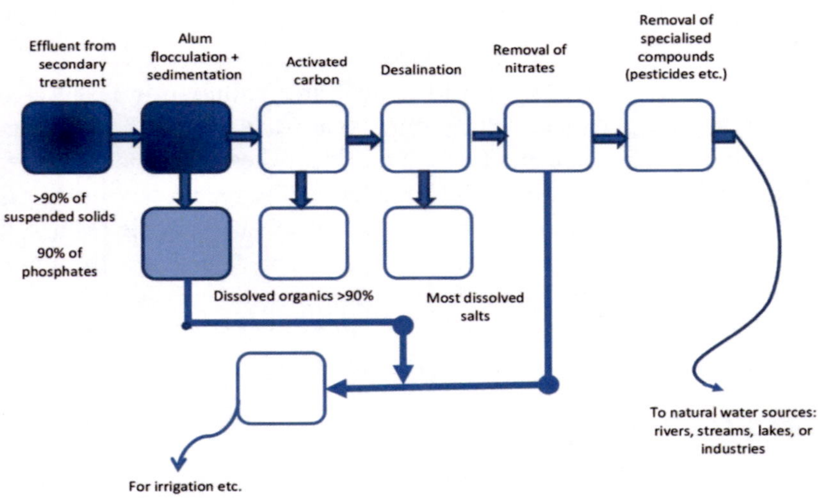

Fig. 3.8: Schematic representation of tertiary treatment of sewage in an effluent treatment plant

Tertiary Treatment Stage

Physico-chemical processes are applied to remove metals, organic solvents, nitrates, phosphates, pathogens, etc. Chemical oxidation of wastewater using strong oxidizing agents, e.g., UV radiation, perchlorate salts, chlorine gas, and ozone gas. The tertiary treatment is followed by the discharge of the treated water into natural sources or used for irrigation or other economic purposes.

Prevention and Control of Water Pollution

In most cases the world over, natural water resources, especially rivers and streams are being polluted because of the dumping of industrial effluents into them. The effluents an industry generates must be treated before they are dumped into a water body. This would greatly control water pollution of the natural sources. Technologies for the treatment of various types of pollutants are well-developed and easily accessible. The acidity and alkalinity of the effluents can be easily brought to neutrality. Metallic salts can be removed by precipitation technique. Colloidal materials present in the effluent can be easily coagulated. Strict implementation of the treatment process greatly helps control water pollution at the very site from where pollutants enter into water bodies.

Thermal pollution can be controlled by cooling the water before it is released from the power plants and drained into water bodies.

Agriculture has become one of the major sources of water pollution due to indiscriminate applications of chemical fertilizers and pesticides. Alternative agricultural practices, such as organic farming, ecological farming, natural farming, etc. offer a solution to prevent and control agriculture-borne pollution. Alternative agriculture is emerging in a big way all over the world, but still, chemical fertilizers and pesticides are posing serious problems. The world's governments have to be serious about the prevention of pollution generated through food production practices.

Bregoli, a by-product of the paper industry akin to sawdust, can clean up oil spills in water bodies. In the case of oil spills, prevention is better than control. The oil import-export business goes on throughout the world. All countries of the world depend on petroleum as a fuel. Foolproof technologies and management must come into effect to avoid oil spills.

There are some techniques available that can remove many common pollutants in industrial effluents. Some are mentioned in Table 3.3.

Table 3.3: Techniques for removal of selective pollutants in industrial effluents

Pollutant	Remediation Technique
Mercury	Mercury-selective ion exchange resin
Ammonia	Ion exchange technique; weak acid cation exchange removes NH_3 in the form of $(NH_4)_2SO_4$
Sodium salts	Reverse osmosis
Phenolics	Polymeric absorbents
Wastewater color	Discoloration: electrolyte decomposition technique
PCBs, dioxins, cyanides, explosives	Photocatalytic reactions (sunlight + catalyst like titanium dioxide)

Watersheds in mountain areas should be covered with dense forests. Tree felling and deforestation in mountain areas should be strictly prohibited. A dense cover of vegetation will most effectively control the problem of the siltation of rivers and water pollution due to turbidity.

Self-Purification of Lotic Ecosystems

Running waters (lotic ecosystems) are exposed to some environmental factors due to which they can purify themselves. The factors that help the natural lotic water bodies, such as rivers, streams, and canals, are physical, biological/biochemical. Water flowing with effluents purifies itself following the distance it flows and time subject to operating physical, chemical, and biological factors (Fig. 3.9).

Physical Factors

The physical factors include sunlight, oxygen, dilution and dispersion, and sedimentation. When a river flows with effluents dissolved in it, they go on diluting and disperse as the water covers more and more distance. The heavier suspended particles keep on settling down in the bottom of the flowing water system following the process of sedimentation. Sunlight is an essential input for the photosynthesizers that help maintain a higher DO level of a water system. Light also induces the growth of aerobic microbes that decompose the organic matter at a faster pace. Flowing water has a greater capacity to dissolve oxygen. A spring dissolves more oxygen than a river. The faster the flow of water the greater the degree of turbulence and higher the DO concentrations in a river. Rivers in the mountains, thus, generally carry more DO. Higher DO concentration means lower BOD and COD values.

Chemical Factors

The chemical processes in the presence of aerobic and anaerobic microbes significantly contribute to the self-purification processes in a river after a certain distance and time. Oxidation of organic matter in the effluents occurs by microbes soon after it is released into a river. In this process, the organic matter is converted into minerals and simple inorganic substances. This process decreases oxygen in the water body which soon is compensated by the growth of photosynthesizers like algae and cyanobacteria. This is how the aerobes aid the process of self-purification. The organic matter also undergoes hydrolysis chemically or by anaerobic microbes. This occurs in the absence of oxygen in which the organic matter gets reduced to gases such as methane and minerals. Thus, anaerobes also contribute to getting rid of the organic matter by promoting reduction processes.

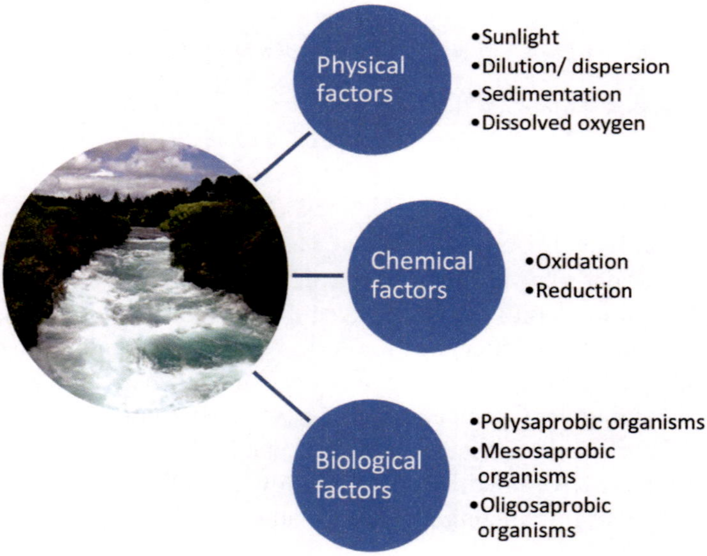

Fig. 3.9: Self-purification process of a lotic water body depending on various factors

Biological factors

The composition of the living organisms in water depends upon the level of pollution, i.e. concentrations of organic and inorganic substances and consequent DO levels. With extreme levels of pollution, the organisms that inhabit the water are polysaprobic type. Moderate to a high level of pollution helps the Mesosaprobic type organisms. The organisms that flourish in the least polluted water are the oligosaprobic ones (Table 3.4).

Table 3.4: Various organisms prospering in different levels of water pollution

State of the water pollution	Organism type	Organisms
Extreme	**Polysaprobic organisms**	
	Algae	*Euglena*, *Phacus,* and diatoms
	Blue green algae (Cyanobacteria)	*Spirulina,Oscillatoria, Microcystis, Phormidium, Anthrospira etc.*
	Animals	*Tubifex, Rotaria, Colpidium, etc.*
Moderate to high	**Mesosaprobic organisms**	
	Algae	*Spirogyra, Cladophora, Uronema,* etc. and diatoms
	Blue green algae (Cyanobacteria)	*Oscillatoria, Spirulina, Phormidium,* etc.
	Animals	*Cloeon, Brachinus, Vorticella, Sphaerium, Coleps,* etc.
Low	**Oligosaprobic organisms**	
	Algae	*Ulothrix, Cladophora, Ulothrix, Bulbochaete,* and diatoms
	Animals	*Vorticella, Planaria, Notholca, etc.*

Various processes involving specific actions of various types of organisms in the presence of various physical factors contribute to bringing flowing waters to a normal condition along with distance and time. However, dumping copious amounts of untreated effluents in the lotic waters at short distances badly affects the self-purifying capacity of the rivers. This is what is the state of rivers in India, including the Ganga and Yamuna, which are polluted to the extent that their self-purification capacity is overreached. This natural power of the lotic systems can be regained by strictly following effluent treatment techniques and avoiding or minimizing pollution loads.

Summary

Water pollution is defined as the mixing of some substances, be they inorganic, organic, radiological, or biological, or a factor, such as heat, which deteriorate water quality to the extent that it either becomes a health hazard or unfit for the intended use. Major sources generating pollutants introduced in water are community wastes, industrial wastes, agriculture, etc. The level of free molecular oxygen in water is referred to as dissolved oxygen (DO). The concentration of DO in aquatic habitats is one of the basic indicators of the quality of water. If the DO content of a water body is recorded below 8.0 mg per liter, it may be considered contaminated. If the same comes down to 4.0 mg per liter, water is considered "highly polluted." Oxygen enters into the water from the atmosphere and through photosynthesis. DO saturation means that water is holding as many O_2 molecules per unit volume as it can in equilibrium. Temperature, pressure, and salinity are the major factors the DO levels of a water body depend upon. BOD, expressed in mg of O_2 per liter of water, is the measure of DO needed by decomposers in the decomposition of organic matter present in water. COD is a measure of oxygen equivalent of the requirements for the complete oxidation of total organic substances, biodegradable as well as non-biodegradable, present in the water. The COD value of the polluted water, thus, is always greater than that of BOD.

Groundwater sources are polluted as a result of seepage from agricultural run-off and through industrial effluents, municipal wastes, sewage channels, and landfills. Overexploitation of groundwater resources for irrigation and drinking water supplies has further aggravated the pollution problem. Oceans and seas are severely polluted by garbage and sewerage of coastal cities, chemical and nuclear wastes, and navigational discharges containing all kinds of pollutants, such as oils, grease, detergents, radioactive wastes, human wastes, etc. Marine pollution by petroleum and petroleum products is one of the gravest global environmental issues. It may be incidental, accidental, or intentional. In the process of pollution, spilled oil is partitioned into the environment in four ways: spreading, evaporation, solubilization, and residual material.

Decreased DO levels affect cold-water fish, like trout and salmon. Not only organic and inorganic pollutants but even hot waters released from some industrial units into natural water bodies also decrease DO concentrations leading to changes in the composition of aquatic species. The presence of some toxic substances, especially pesticides, and heavy metals such as lead, mercury, arsenic, etc. leads to eliminate several aquatic organisms in water bodies. The phenomenon of biological magnification especially applies in case of the non-degradable and persistent pollutants such as pesticides. The process of nutrient enrichment of a water body leading to changes in species composition, the density of populations, and loss of biodiversity is known as eutrophication.

Human health is highly susceptible to water pollution. Most of the diseases humanity suffers from the world over are attributable to the use of polluted water. The use of water polluted by sewage causes what we often refer to as water-borne diseases, such as amoebiasis, diarrhea, jaundice, typhoid, cholera, etc. Water contaminated by cadmium causes Itai Itai disease, also called ouch-ouch disease, a painful disease of bones and joints. The oft-quoted disease caused by the consumption of fish from a water source contaminated with mercury is the Minamata disease, a neurological syndrome, reported in Japan in 1956. Water polluted with nitrates is especially fatal for infants causing the blue baby syndrome. Upon reaction with hemoglobin in the blood, nitrates form methemoglobin which does not transport oxygen to body tissues. This health problem is called methemoglobinemia. Overexploitation of groundwater resources may result in arsenic from soil and rock sources enter into and poison the water. The use of arsenic-contaminated water leads to what is known as black foot disease. Overexploitation of groundwater also leads to water pollution by excessive fluorides. Fluoride excess in potable water causes what is called skeletal fluorosis characterized by teeth deformity, stiff and painful joints, and hardening of bones. Non-skeletal fluorosis results in gastrointestinal and neurological problems.

Sewage is one of the prime management systems of urban life the world over. Technologies for the treatment of various types of pollutants are well-developed and easily accessible. The acidity and alkalinity of the effluents can be easily brought to neutrality. Metallic salts can be removed by precipitation technique. Colloidal materials present in the effluent can be easily coagulated. Thermal pollution can be controlled by cooling the water before it is released from the power plants and drained into water bodies. Bregoli, a by-product of the paper industry akin to sawdust, can clean up oil spills in water bodies. In the case of oil spills, prevention is always better than control.

Running waters (lotic ecosystems) are exposed to some environmental factors due to which they can purify themselves. This natural power of the lotic systems can be regained by strictly following effluent treatment techniques and avoiding or minimizing pollution loads.

Exercises

A. Multiple Choice Questions

1. What is the likely effect of NO_3 if its concentration exceeds 100 mg per liter of drinking water?

   a. Stone in gallbladder
   b. Methemoglobinemia
   c. Scurvy
   d. Diabetes

2. Which of these is a correct statement?

   a. Water with a DO content of 8 mg/l will only support worms, bacteria, and other decomposers.

   b. Water with a DO content of 8 mg/l will support desirable forms of aquatic life.

   c. Both (a) and (b) above

   d. None of the above

3. The pollution load of water in a water body due to sewage or other organic wastes is measured in terms of

   a. DO b. COD

   c. BOD d. TSS

4. Which of the following is an incorrect statement?

   a. A large amount of sewage in water generates high BOD and low levels of DO

   b. BOD value is always less than that of COD.

   c. The rate of decomposition is slower in warm water than in cold water.

   d. Cool water dissolves more oxygen than hot water.

5. In its aging, a natural lake advances from

   a. New to old aquatic ecosystem b. Disclimax to climax

   c. Eutrophic to oligotrophic d. Oligotrophic to eutrophic

6. Which is/ are the most associated with the eutrophication of a lake?

   a. N b. P

   c. K d. N and P

7. Which organism in drinking water causes amoebic dysentery?

   a. *Entamoeba histolytica* b. *Vibrio cholera*

   c. *Salmonella typhi* d. *Taenia saginata*

8. Which element was responsible for the Minamata disease when this element entered the food chain in the Minamata Bay of Japan?

   a. Hg b. As

   c. Cd d. Pb

9. Water in a lake was analyzed and the concentration of polychlorinated biphenyls (PCB) in it was found to be 0.000002 ppm. When the concentration of PCB was determined in phytoplanktons, zooplanktons, small fish, large fish, and in the eggs of fish-eating gulls, it was found to be gradually increasing, with as high as 124 ppm in seagull eggs. This is an example of

   a. Cyclomorphosis b. Biological magnification

   c. Biodegradation d. Eutrophication

10. In a lake, a rise in temperature may lead to
    a. The disappearance of blue-green algae and the dominance of diatoms
    b. Increase in BOD and consequent depletion of DO
    c. Increase in DO and a corresponding decrease in BOD
    d. Increase in both DO and BOD
11. What is/ are the main contributor(s) to excessive algal blooms in mangrove ecosystems?
    a. Nutrient enrichment
    b. Temperature increases
    c. None of (a) and (b) above
    d. Both (a) and (b) above
12. By causing O_2 depletion and releasing toxins, algae blooms kill deep-dwelling fish. An example was the collapse of the cod fishery in the early 1990s in the
    a. Minamata Bay
    b. Arabian Sea
    c. Baltic Sea
    d. Dead Sea
13. The tertiary sewage treatment is
    a. Physical treatment
    b. Biological treatment
    c. Biophysical treatment
    d. Chemical treatment
14. A high BOD of a sample is an indicator of
    a. No microbial pollution
    b. Less microbial pollution
    c. Intense microbial pollution
    d. All of the above
15. COD is a better indication of water pollution because
    a. It is very simple
    b. It is not costly
    c. It is less time consuming
    d. It measures water pollution even by non-biodegradable wastes
16. The amount of DO present in water depends on
    a. Surface turbulence
    b. Photosynthetic activity
    c. Decomposition of organic matter
    d. All of the above

17. Chronic exposure to arsenic causes
    a. Black foot disease
    b. Grey baby syndrome
    c. Hey baby syndrome
    d. None of the above
18. An accidental discharge of petroleum in oceans and estuaries is known as
    a. Oil spill
    b. Organic spill
    c. Waste spill
    d. Black pollution
19. Epiphytic lichens could be used as indicators of
    a. NO_2 pollution
    b. SO_2 pollution
    c. H_2S pollution
    d. O_3 pollution
20. Water contaminated with cadmium can cause
    a. Red color disease
    b. Itai-Itai disease or ouch-ouch disease
    c. Blue baby syndrome
    d. Black foot disease

Answers: 1-**b**, 2-**b**, 3-**c**, 4-**c**, 5-**d**, 6-**d**, 7-**a**, 8-**a**, 9-**b**, 10-**b**, 11-**a**, 12-**b**, 13-**d**, 14-**c**, 15-**d**, 16-**d**, 17-**a**, 18-**a**, 19-**b**, 20-**b**

B. Short-Answer Questions

1. What is water pollution?
2. Mention point and non-point sources of water pollution.
3. Name the biological organisms that cause water pollution.
4. What is the importance of the DO value of water bodies?
5. Differentiate between BOD and COD.
6. What health problem is caused by drinking arsenic-contaminated water?
7. What causes Itai-Itai disease?
8. What is the cause of Minamata disease?
9. Name the health problem in infants caused by drinking water polluted by nitrates.
10. What do you understand by biological magnification?

C. Long-Answer Questions

1. What are the main causes of water pollution in our contemporary world? Describe how extensively various development sectors are spoiling water quality and natural water bodies.
2. Describe various effects of water pollution on: i) aquatic life, ii) plants, and iii) human health.
3. Suggest various ways and means of preventing and controlling water pollution.
4. Explain the processes of sewage management with the help of a representative diagram.
5. Write short notes on the following:

   (a) Groundwater Pollution

   (b) Marine Pollution

   (c) Eutrophication

Suggested Readings

Carpenter, S. R. 1981. Submersed vegetation: an internal factor in lake ecosystem succession. The *American Naturalist* 118: 372-383.

Chislock, M. F., Doster, E., Zitomer, R. A. and Wilson, A. E. 2013. Eutrophication: Consequences, and controls in aquatic ecosystems. *Nature Education Knowledge* 4(4): 10.

Freedman, B. 1989. Environmental Ecology: The Impacts of Population and Other Stresses on Ecosystem Structure and Function. San Diego: Academic Press Inc. 424 pp.

Turner, A. M. and Chislock, M. F. 2010. Blinded by the stink: nutrient enrichment impairs the perception of predation risk by freshwater snails. *Ecology Applications* 20: 2089-2095.

Web Resources

https://byjus.com/biology/water-pollution-control/

https://wwf.panda.org/knowledge_hub/teacher_resources/webfieldtrips/water_pollution/

https://www.embibe.com/exams/water-pollution/

https://www.environmentalpollutioncenters.org/water/

https://www.fondriest.com/environmental-measurements/parameters/water-quality/dissolved-oxygen/

https://www.nationalgeographic.com/environment/article/freshwater-pollution

https://www.nature.com/scitable/knowledge/library/eutrophication-causes-consequences-and-controls-in-aquatic-102364466/

https://www.nrdc.org/stories/water-pollution-everything-you-need-know

https://www.sciencedirect.com/topics/social-sciences/water-pollution

4

Soil Pollution

Henry & Co./ Unsplash

"*A nation that destroys its soils, destroys itself.*"

– Franklin D. Roosevelt

Life on Earth is rooted in soil. Soil is not just a substratum anchoring the plants and giving a base for all terrestrial animals. It is the largest terrestrial ecosystem, and one of the most wonderful ecosystems on Earth. It inhabits most microorganisms. Microbial populations flourishing within the soil have greater biomass than all the human beings, elephants, and whales put together. It is due to the microorganisms that the soil constitutes the most active ecosystem on Earth.

The health of the soil is crucial for the health of all the plants growing on it and for all the animals dependent on it for their inhabitation and nourishment. Soil health is also critical for the well-being of numerous microorganisms, detritivores, and several other organisms. Soil is composed of all essential nutrients/ minerals, air, water, and organisms in a proportion that strikes its ecological balance and maintains its structure and functions revealed through its productive potential.

Soil, like all other ecosystems, is itself a dynamic ecosystem. Its composition keeps changing but natural inbuilt mechanisms help strike a balance and recover natural composition. But when its composition changes to a considerable extent, especially due to external factors, the original characteristics and functional vigor of the soil undergo undesirable alterations. This state of the soil is said to be soil pollution. Soil pollution, thus, is defined as an undesirable change in the natural physical, chemical and biological composition of the soil.

Sources of Soil Pollution

Soil pollutants can be natural as well as of anthropogenic origin. Natural sources are not always a nuisance, the anthropogenic ones always are.

Natural Sources

Plant Residues and Chemicals

All green plants add carbon content to the soil that serves as a culture medium for numerous types of microorganisms and other organisms, builds up humus, maintains soil moisture, and enhances soil fertility. However, some plants also leave some residues and release some chemicals into soils that act as pollutants. The residues and chemicals might affect soil composition and/ or be detrimental to soil organisms.

Forest fires and burning of crop residues add unwanted compounds that create unfavorable conditions in the soil environment leading to the wiping out of soil microflora and reduced plant growth. Carbon monoxide (CO), nitrogen oxide (NO), sulfur dioxide (SO_2), hydrocarbons, etc. are all detrimental to soil life. High concentrations of CO are toxic to the roots of plants, microorganisms, and higher forms of life in the soil. NO, and NO_2 react with soil moisture and form nitrous and nitric acids which bring down soil pH. In the presence of soil moisture, SO_2 forms sulfuric acid which also increases soil acidity. Accumulation of these acids and consequent acidification of the soil leads to the wiping out of soil organisms and gradually turns the soil unfertile and uninhabitable for plants.

Many plant species release allelochemicals from their roots, shoots, and leaves. The allelochemicals the allelopathic species release in the soil environment are deleterious for the other species (Rastogi and Singh 2019, Singh 2020). Thus, some chemicals may affect the growth of microorganisms and other plant species. These chemicals persist in the soil environment, sometimes for a longer period, and affect the growth and survival of many plant species. However, the allelochemicals are biodegradable and ultimately they cease to affect the soil environment and plant growth.

Animal Wastes

Animal wastes such as excreta, abattoir effluents, carcasses, dead bodies, etc. continuously dumped into the soil in a limited area spoil the soil environment by causing hypertonic conditions. Such conditions are extremely detrimental to the growth of plants and also spoil the environment outside the soil. Animal wastes are biodegradable and should not be regarded as pollutants per se. Upon decomposition, naturally, they convert into simple substances and recycle into various environmental components. The problem of soil pollution owing to animal wastes arises only due to anthropogenic management. Despite being natural pollutants, thus, animal wastes are more akin to pollutants of anthropogenic origin. In abattoirs etc., such human activities causing widespread pollution and diseases, are utterly inhuman and unethical activities causing incalculable harm to ecosystems and miseries to life.

Anthropogenic Sources

All human lifestyles are generating pollutants and contribute to soil pollution in one way or the other.

Domestic Wastes

Domestic wastes are the major cause of soil pollution. These wastes are mostly solid or semi-solid forms called solid wastes. The waste is generated in copious amounts from human settlements: towns and cities, shopping centers, hospitals, restaurants, hotels, education institutes, office establishments, rural industries, etc. These wastes keep accumulating in the neighborhood of human establishments spoiling soils and gradually turning them into unproductive deserts.

The other most common use in households is that of detergents. These detergents flow freely along with domestic effluents. Accumulated in soils the detergents gradually increase alkalinity and phosphate contents leading to a decline in microbial populations and a continuous decrease in soil fertility.

Industrial Wastes

Beginning in the middle of the eighteenth century, the Industrial Age gradually changed the entire global scenario including human lifestyles and socioeconomic activities. Now almost all the Earth's ecosystems are modified ecosystems and environmental pollution is pervading all over the globe. Industries generate all kinds of pollutants—solids, semi-solids, liquids, and gases, toxic and non-toxic, metallic and non-metallic, persisting and short-living, life-annihilating as well as neutral to life.

Industrial effluents are often laden with deleterious chemicals and heavy metals like copper, zinc, chromium, nickel, iron, etc. The toxic chemicals like cyanides and heavy metals gradually seep through the soil and make entry into groundwater. The tannery industry in Kanpur city of India has chromium in the toxic effluent it generates which is causing problems for the crop fields and farmers. Industrial wastewater flowing from a paper mill located in Lalkuwan in the Tarai area of Uttarakhand State of India has been reported to spoil agricultural lands and crops (Rastogi et al. 2011). Groundwater in several areas has been reported contaminated with heavy metals (Parmar and Singh 2015, Negi et al. 2019). Some toxic chemicals reach the soil through mining operations. Industrial effluents gradually deplete life within the soil and eventually convert the fertile lands into mooring landscapes.

Acid rain is another 'non-point' type of pollution borne out of industrial pollution. Oxides of sulfur and nitrogen upon reaction with rainwater form secondary pollutants, sulfuric acid, and nitric acid, respectively. Acid rain lowers soil pH and impoverishes soil fertility.

Military activities and wars are also the major causes of the metal pollution of the soil. This is one of the least examined causes of soil pollution. The other least known cause of soil pollution is urban vehicles' corrosion which contributes to metal pollution of the soil. Both weapons of war and vehicles are robust products of the industry.

Agrochemicals

Modern agriculture developed around the so-called Green Revolution technology is heavily dependent on agrochemicals, mainly chemical or synthetic fertilizers and pesticides. This agriculture hardly recognizes soil as an ecosystem, but only manipulates soil fertility by applying chemical fertilizers, especially nitrogenous fertilizers. NPK, bone meal and superphosphates and mined fertilizers are the other types of fertilizers applied in cultivated lands. The Green Revolution promotes monocultures that do not have resilience enough to resist the onslaught of plant pathogens. Therefore, pesticides and herbicides are also used as essential inputs in agriculture.

Excess doses of chemical fertilizers create a nutrient imbalance in the soil. Excessive accumulation of nitrates in the soil is linked with widespread repercussions of soil pollution. Excess phosphates cause adverse effects on plant growth. Indiscriminate applications of chemical fertilizers change soil structure and nutrient composition and may even knock down soil flora and fauna.

Pesticides are used in cultivated crops to protect them against a variety of insects and disease-causing organisms. Insecticides, fungicides, nematicides, rodenticides, etc. are the various types of pesticides employed in agriculture. These pesticides poison the soil with toxic chemicals and heavy metals and knock down soil flora and fauna as well as useful pollinators and, thus, create an ecological imbalance within and outside the soil ecosystem. Herbicides kill not only unwanted plants (weeds) but also useful plants and, thus, destroy plant diversity.

Radioactive Wastes

Radioactivity, despite its incurable dangers, has become an inevitable evil of the contemporary human world. Nuclear explosions, nuclear tests, nuclear reactors, laboratory experiments involving radioactive isotopes, etc. leave behind what could be the worst kind of pollutants: the nuclear wastes. From the contaminated soil, radioactive isotopes enter into plants growing on it, and from the plants, they enter into the human body through the food chain. In the human system, they play an extremely detrimental role.

The worst aspect of the problem is that these pollutants in the soil cannot be tackled or neutralized. Further, there is no cure for the destructive effects of radioactive pollutants if they enter human systems. Radioactivity of the elements in the contaminated soil persists for a pretty long time. The harmful effects of radioactive wastes reduce only with time.

Effects of Soil Pollution

Soil pollution is not an exclusive occurrence in itself. It has widespread implications for water, air, plants, animals, and human systems. All terrestrial life is dependent on soil. If soil is healthy, terrestrial organisms will also stay healthy. If soil is sick with pollutants, all terrestrial beings would have to bear the implications. Soil sickness has far-reaching implications. Ultimately, death creeps over sick soil. Dead soils turn into unproductive deserts (Fig. 4.1). History of the world would reveal that the civilizations that did not care for their soil vanished from the face of the Earth. Our socioeconomic progress is also rooted in soil. Unhealthy soil adversely affects socioeconomic progress. All the ill effects of soil pollution are linked with each other (Fig. 4.2).

Fig. 4.1: Death creeping on the polluted land: Soil pollution eventually leads to an unproductive desert (image credit: Pexels-Cottonbro Studio)

Pollution of the soil seriously hits the biogeochemical cycles, not only the sedimentary types but also the gaseous ones. Poor soil grows poor crops. The productivity of cultivated crops and biomass of natural vegetation growing on polluted soils are poor. Soil pollution, eventually, affects the functioning of the soil ecosystem.

Many sensitive plant species cease to grow in polluted soils. The polluted soils, thus, lead to biodiversity loss and may push many valuable species toward extinction. Perhaps the worst implication of polluted soils is that they may curtail the photosynthetic efficiencies of the ecosystems leading to reduced carbon sequestration, accelerated carbon emission, impoverished carbon storage of ecosystems, and subsequently global warming and climate change. None of the effects of soil pollution is isolated but they have a triggering effect and, ultimately, the entire biosphere comes into the grip of the adverse chain reactions.

Toxic chemicals and toxic and potentially toxic pollutants are eventually absorbed by plants growing on contaminated soils and through food enter into human bodies. Contaminated soils, thus, spread several kinds of metabolic disorders among human beings. The soil pollutants cause the same human health problems as the air and water pollutants do, as discussed in other chapters.

The use of urban wastewater, sewage sludge, and livestock manure in agriculture is often overemphasized in lectures on waste management. However, use of the untreated wastes may pose a serious risk as these may serve as a source of anti-microbial substances resulting in the presence of anti-microbial resistant bacteria in soils (Kuppusamy et al. 2018). Looking at this state of waste generation and consequent soil pollution, we should revisit O'Neill (2014) estimate that antimicrobial-resistant infections might become the dominant cause of death in the world by 2050.

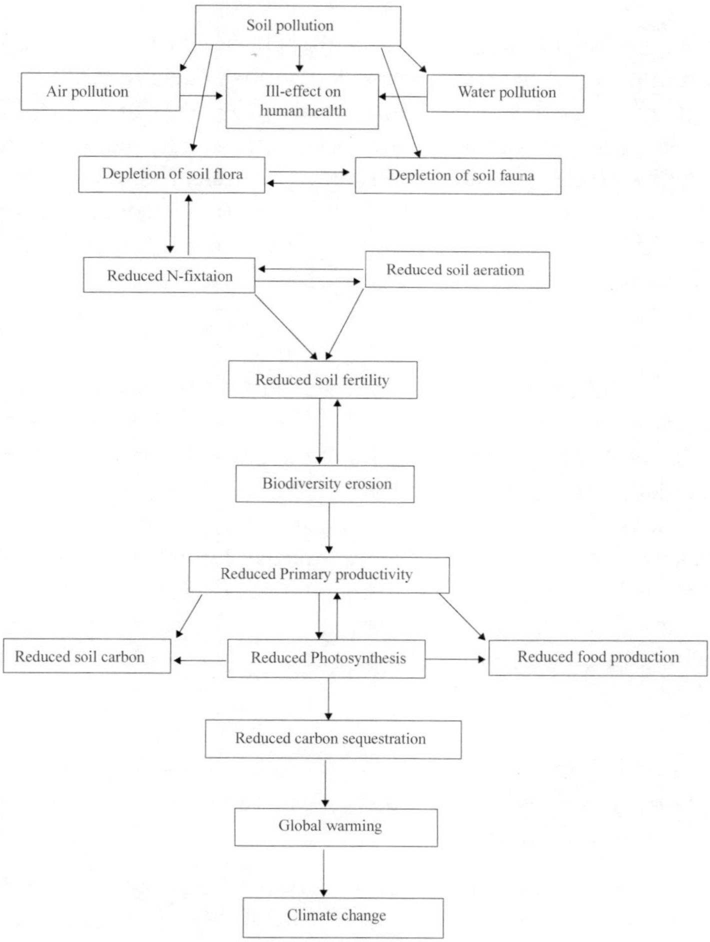

Fig. 4.2: The triggering effects of soil pollution

Prevention and Control of Soil Pollution

Domestic Wastes

Domestic sources of soil pollution can be managed by managing sanitary landfills. In this process, even electricity can be generated by using the fuel source coming out of the leachate of the garbage. Organic wastes can also be converted into compost. Biogas can also be produced by fermenting organic matter in specific digesters. Such a gas can be used for cooking purposes. Sewage should not be dumped into soils. The biological treatment of sewage helps turn it into manure.

Industrial Wastes

Industrial wastes should not be dumped into soils. These can be selectively used and converted into useful products, such as compost, fuels, etc. It is the untreated industrial effluents that contaminate soils after they are released as such into rivers, streams, or standing water bodies. Effluents generated in industrial units must be properly treated before they are released into natural water sources. Acid rains that also pollute soils can also be controlled by minimizing pollution due to emissions of carbon, sulfur, and nitrogen oxides from the industrial sector.

Agrochemicals

Ecologically sound and sustainable agricultural systems do not use chemical or synthetic fertilizers and pesticides or herbicides. If such agricultural systems prevail, much of the problem of soil contamination through the agriculture sector would be avoided. Alternative agriculture, such as ecological agriculture, organic agriculture, natural farming, etc. is increasingly becoming popular as it avoids external chemical inputs and uses techniques and inputs that are eco-friendly and health-ameliorating. Biodiversity-rich agroecosystems are more resilient and sustainable and insects and plant pathogens in these ecosystems can be managed without the application of pesticides.

Radioactive Wastes

The use of radioactive isotopes needs to be minimized. Radioactive wastes should be disposed of with utmost care. It must be ensured that the radioactive substances do not enter into food chains.

Soil pollution is certainly a hidden danger for life on the planet as well as for human health and future progress. It is one of the gravest as well as the least addressed problems of our times. Many of the ecological, and environmental problems, including the most serious of climate change, as well as socioeconomic stresses our contemporary world faces are rooted in soil pollution. Soil has an inherent capacity for filtering, buffering, and attenuation. But these capacities are finite and at many sites of the world seem to have overreached. As soil health is compromised, food and water quality become worse. Depolluting soil and maintaining soil health must be one of our top priorities.

Summary

Soil pollution is defined as an undesirable change in the natural physical, chemical and biological composition of the soil. Soil pollutants can be natural as well as of anthropogenic origin. The natural sources may be plant residues and chemicals released that might affect soil composition and/ or be detrimental to soil organisms. Forest fires and burning of crop residues add unwanted compounds that create unfavorable conditions in the soil environment leading to the wiping out of soil microflora and reduced plant growth. Many plant species release allelochemicals from their roots, shoots, and leaves which could be deleterious for the other species. Animal wastes such as excreta, abattoir effluents, carcasses, dead bodies, etc. continuously dumped into the soil in a limited area spoil the soil environment by causing hypertonic conditions. Anthropogenic factors include domestic wastes (e.g., detergents), industrial wastes (e.g., heavy metals, toxic chemicals), agrochemicals (pesticides and chemical fertilizers), and radioactive wastes.

Soil pollution has widespread implications for water, air, plants, animals, and human systems. Pollution of the soil seriously hits the biogeochemical cycles. The productivity of cultivated crops and biomass of natural vegetation growing on polluted soils are poor. Soil pollution, eventually, affects the functioning of the soil ecosystem. Contaminated soils spread several kinds of metabolic disorders among human beings.

Organic domestic wastes should be converted into compost. Biogas can also be produced by fermenting organic matter in specific digesters. Industrial wastes should not be dumped into soils. These can be selectively used and converted into useful products, such as compost, fuels, etc. Effluents generated in industrial units must be properly treated before they are released into natural water sources. Acid rains that also pollute soils can also be controlled by minimizing pollution due to emissions of carbon, sulfur, and nitrogen oxides from the industrial sector. Ecologically sound and sustainable agricultural systems not dependent on chemical or synthetic fertilizers, pesticides, and herbicides would prevent soil pollution due to the agriculture sector. Radioactive wastes should be disposed of with utmost care. Depolluting soil and maintaining soil health must be one of our top priorities.

Exercises

A. Multiple Choice Questions

1. Which of the following can cause soil pollution?

   a. Heavy metals b. Radioactive elements

   c. Herbicides d. All of the above

2. Which of the following sectors is less likely to cause soil pollution?

   a. Agriculture b. Livestock

   c. Industry d. Forest

3. Which of the following is a false statement?

   a. Some toxic chemicals reach the soil through mining operations.

   b. Nitrogenous fertilizers pollute the soil, but phosphorus and potassium do not.

   c. Pesticides may add to heavy metal pollution of the soil.

   d. Soil pollution may cause air and water pollution.

4. Which of the following can cause soil pollution?

   a. Allelopathic plants b. Earthworms

   c. Nitrogen fixers d. Symbionts

5. What revolutionary change in the 1960s became one of the worst causes of soil pollution?

   a. Green Revolution b. White Revolution

   c. Blue Revolution d. Industrial Revolution

6. What kind of farming is not safe and healthy for the soil?

   a. Green Revolution or modern farming

   b. Organic farming

   c. Ecological or sustainable farming

   d. Natural farming

7. If the soil is sick, i.e., polluted

   a. Plants growing on it would hardly be affected.

   b. Soil flora and fauna won't be affected.

c. Plants, animals, and human beings dependent on it would not stay healthy.

d. Its productivity would decline, but it would not turn into a desert.

8. If this pollutant accumulates in the soil, it may flow to water bodies and may cause a blue baby syndrome in the children who drink the water contaminated by it. This pollutant is

a. Lead
b. Nitrates
c. Arsenic
d. Iron

9. Accumulated in excess, which of the following metals would be a soil pollutant?

a. Iron
b. Magnesium
c. Copper
d. All of the above

10. Salination of the soil is caused by an increase in the concentration of

a. Soluble salts
b. Acids
c. Alkali
d. Nitrates

11. Which of the following is a true statement?

a. Soil pollution leads to declined crop productivity, but it has no linkage with climate change.

b. Soil pollutants can change soil structure but not the soil composition.

c. The harmful effects of the radioactive wastes reduce only with time.

d. None of the above

12. Which of the soil pollutants are organic?

a. Heavy metals

b. Allelochemicals

c. Carbon monoxide gas after forest fire

d. Sodium salts

13. Which of the following is a natural source of soil pollution?

a. Forest fires
b. Allelopathic species
c. Both of the above
d. None of the above

14. What cropping system is not healthy for the soil?
    a. Monocultures
    b. Agro-biodiversity based cultures
    c. Agroforestry
    d. Silviculture
15. The soil's capacities to mitigate its pollution are
    a. Finite
    b. Infinite
    c. Dependent on a site/ geographical region
    d. Infinite if managed properly
16. What is affected by soil pollution?
    a. Carbon sequestration
    b. Global warming
    c. Soil ecosystem functioning
    d. All of the above
17. What can the corrosion of urban vehicles add to soil pollution?
    a. Metals
    b. PAN
    c. Ozone
    d. Varnish
18. Military activities and wars are big but unnoticed sources of soil pollution. Due to these, the soil is contaminated by
    a. Heavy metals
    b. Discarded dresses
    c. Boots
    d. None of the above
19. Mycoremediation of the contaminated soil involves
    a. Algae for the extraction of pesticides
    b. Fungi for the accumulation of heavy metal contaminants
    c. nanoparticles for the absorption of toxins
    d. Microorganisms for the absorption of heavy metals
20. Phytoremediation means the removal of metals from the soil using
    a. Certain microbes
    b. Fungi
    c. Selective plant species
    d. All of the above

Answers: 1-**d**, 2-**d**, 3-**a**, 4-**a**, 5-**a**, 6-**a**, 7-**c**, 8-**b**, 9-**d**, 10-**a**, 11-**c**, 12-**b**, 13-**c**, 14-**a**, 15-**a**, 16-**d**, 17-**a**, 18-**a**, 19-**b**, 20-**c**

B. Short-Answer Questions

1. What is soil pollution?
2. What are the natural causes of soil pollution?
3. How do radioactive elements cause soil pollution?
4. What pollutants are there in the agrochemicals?
5. How do domestic wastes cause soil pollution?
6. How are industrial wastes a source of soil pollution?
7. Why do monocultures in agriculture essentially need the use of pesticides?
8. Repeated applications of herbicides lead to biodiversity erosion. How?
9. How does soil pollution affect human health?
10. How does soil pollution simultaneously affect air and water pollution?

C. Long-Answer Questions

1. Describe the anthropogenic sources of soil pollution.
2. Explain with the help of a flow diagram the interlinked and widespread implications of soil pollution.
3. Management of the soils can be pivotal in mitigating climate change. Explain with reasoning.
4. How would you prevent and control soil pollution? Explain giving examples.
5. Write short notes on the following:

   (a) Green Revolution and soil pollution

   (b) Domestic wastes as soil pollutants

   (c) Industrial waste and soil pollution

Suggested Readings

Kuppusamy, S., Kakarla, D., Venkatswarlu, K., Megharaj, M., Yoon, Y. E. and Lee, Y. B. 2018. Veterinary antibiotics (VAs) contamination as a global agro-ecological issue: a critical view. *Agric. Ecosyst. Environ.* 257: 47-49. doi: 10.1016/j.agee.2018.01.026

Negi, S., Singh, Vir, Melkania, U. and Rai, J.P.N. 2019. Estimation of heavy metal pollution index for groundwater around Integrated Industrial Estate of Tarai Region. *Environment and Ecology* 37 (IB): 429-435.

O'Neill, J. 2014. *Antimicrobial Resistance: Tackling a Crisis for the Health and Wealth of Nations*. London: Review on Antimicrobial Resistance.

Parmar, S. and Singh, Vir. 2015. Phytoremediation approaches for heavy metal pollution: A review. *Journal of Plant Science and Research*, 2(2): 1-7.

Rastogi, A., Seth, P. and Singh, Vir. 2011. *Pulp and paper Industry and Environmental Disater: Death of a Stream and Heavy Metal Accumulation in Wheat Crop in Himalayan Foothills.* Lambert Academic Publishing, Germany. 138pp.

Rastogi, A., Singh, Vir and Arunachalam, A. 2019. Allelopathic effects of aqueous leaf extract of *Jatropha curcas* L. on food crops in the Himalayan foothills. *Journal of Emerging Technologies and Innovative Research* 6 (6): 866-875.

Singh, Vir. 2020. *Environmental Plant Physiology: Botanical Strategies for a Climate Smart Planet*. London: CRC Press (Taylor and Francis) 320pp.

Web Resources

http://osou.ac.in/eresources/Soil%20Pollution.pdf
https://byjus.com/chemistry/soil-pollution/
https://en.wikipedia.org/wiki/Soil_contamination
https://www.speakingtree.in/article/an-amazing-world-beneath-your-feet

5

Noise Pollution

Andrea Peacquadio/ Pexels

"Our culture associates noise with power and progress...Physically, "noise" is wasted power for it represents wasted sound (that delivers little useful information), and wasted energy (because electromechanical generation emits heat). But psychologically, we perceive noise as proportional to power and therefore enviable." – Julia Corbett

A sound is a form or energy. Therefore, it has to have an impact on life in one way or the other. Most of the organisms of Kingdom Animalia communicate with each other creating a peculiar sound. Creating sound is inevitable for performing various life activities. Every animal species has the characteristic sound it reflects through its behavior. Whatever all the animals communicate through creating specific sounds is all-natural. Nature, as such is a stupendous repository of a diversity of sounds, all being specific characteristics of life.

Human behavior, however, is highly evolved in terms of creating a variety of sounds. Communication, entertainment, and many behavioral activities are expressed through creating distinctive sounds. There is a difference between natural sounds and the ones human beings create or can create. Humans often generate sounds, not just in a natural manner, but also with various technical aids, that often become unpleasant what we call noise. Creating noise, or noise pollution, seems to be an exclusive characteristic of the human species. Of course, there could be some natural sounds that might occasionally cause havoc to life. Examples are the sounds generated at the time of the volcanic eruption, during an earthquake or tsunami, and when there is a thunderstorm. These sounds, however, are unusual.

Sound pollution, also known as noise pollution, is the propagation of sound with harmful implications for human beings and animals. In other words, noise is undesirable and unwanted sound that becomes an irritant and a source of stress and often of some diseases. Noise pollution sometimes is distinguished based on the place or site it is generated at. Undesirable sound is an inevitable 'product' of the Industrial Age we are creeping through since the middle of the 18th century. Noise pollution, in essence, has become a permanent and unavoidable characteristic of our world.

With sound pollution becoming a common feature of the world, its serious repercussions are also coming to the fore. The phenomenon of this species-specific (human-generated) problem needs to be understood.

Sources of Noise Pollution

The contemporary world is surviving in a sort of technosphere (Singh 2014). Our lives are overwhelmingly dependent on technology. The creation of sound pollution is one of the basic characteristics of the technological devices we have to inevitably rely on (Fig. 5.1).

Fig. 5.1: Contemporary civilization is cursed to live in a technosphere full of splitting noise pollution (image by Kaique Rocha from Pexels)

Let us ponder over some technological developments that cannot work without creating sound in amounts that can be dangerous to life:

- Transport Sector is the most notorious one for creating noise pollution. Living continuously amid the noise the motor vehicles, airplanes, trains and other sources of transport generate is like living in a hell of demonic sounds. Those living near highways, railway tracks, and airports find it extremely distressing and stressful.
- Industrial plants come at number two as far as the generation of sound pollution is concerned. Factory equipment, generators, grinding machines, sirens, etc. are the many industry-related sources of sound pollution. Textile mills, engineering establishments, printing presses, etc. are the major sources of sound pollution in some places.
- Construction machinery, such as bulldozers, road rollers, pneumatic drilling machines, pile driving equipment, stone grinders, etc. are the major sources of sound pollution especially at and near construction sites.
- Defense machinery, equipment, and vehicles, such as fighter planes, artillery, tanks, rocket launching, explosions, practice firings, etc. are other sources of sound pollution. A world living constantly amid the fear of wars and terrorism has to bear the brunt of sound pollution in some of its strategic areas.
- Domestic gadgets such as desert coolers, air conditioners, vacuum cleaners, exhaust fans, mixers, etc. are the sources of noise pollution to be experienced in almost all households.
- Entertainment sources, especially television, radio, tape recorders, and nowadays even mobile phones, are yet another sources of sound pollution almost all families have to encounter in our contemporary world.
- Public address systems occasionally emerge as a notorious source of pollution. The noise pollution created by high-volume loudspeakers on special occasions has implications for social, cultural, and national values.
- Special social-cultural events such as marriages and some festivals also make people exposed to noise pollution created by musical instruments, loudspeakers, disc jockeys (DJs), use of crackers, etc.

Measuring the Sound

Sound propagates in pressure waves and imparts its effect on our eardrums. To travel, the sound needs a medium – solid, liquid, or gaseous. Frequency and intensity are the two very important characteristics of a sound. Frequency (or pitch), a measure of how low or how high a sound is, is measured in terms of waves per second or cycles per second. The intensity (or loudness) of a sound wave is the average per unit area at which energy is transferred by the wave onto the surface, expressed as weber per square meter, Wm^{-2}. Sound measurement has been devised on a logarithmic, rather than linear, scale with a reference to 10 and referred to as **decibel** (*deci* means 10 and *bel* added to recognize the work of Alexander Graham Bell). Abbreviated as dB, decibel measures a sound intensity by a ratio of loudness with reference to the softest audible sound. With ambient sound intensity equal to the reference intensity, the noise level is considered equal to 0 dB.

The range of the noise level can extend from zero to 140 dB (Table 5.1). Beyond 120 dB, the noise causes physical discomfort. Normal conversation sound ranges between 35 dB and 60 dB. Impairment of hearing may be caused if one is exposed to 80 dB or more for a longer period. If you have prolonged exposure to sound intensity of 120 dB, you would feel it to be painful. Continuous exposure to the noisy environment in some factories can make the workers lose their ability to hear soft sounds.

The intensity of sound goes on diminishing as we go away from the point or site where the sound pollution is generated. Generally, the decibel level diminishes at a rate of 6 dB for each 20 meters distance from the source of sound pollution provided there is no reflecting surface in the vicinity. Thus, if a sound measures 140 dB at the source, it would be measured at 134 dB at 20 meters, 128 dB at 40 meters, 122 dB at 60 meters, and so forth.

Table 5.1: Sound level and sources

Sound range (dB)	Sound characteristic	Sources
30	Faint	Whispering, quiet library
40-50	Moderate	Quiet room
60-80	Very loud	Vacuum cleaner, busy street
90	Extremely loud	Lawnmower, truck traffic
120-140	Painful	Jet plane take off, amplified rock music

Effects of Noise Pollution

Noise pollution puts the people exposed to it under stress. It affects human health in various ways. The effects of the undesirable and unpleasant sounds depend on their intensity, frequency, and periodicity.

Absolute Silence

What would be the effect of absolute silence on human health? Absolute silence or a sound level of zero dB, for a prolonged time, is as undesirable as noise pollution. Continuous exposure to absolute silence makes one quite uncomfortable. The pounding of the heart is felt. Even though there is no sound, it feels like something is ringing in the ears. A soundless environment triggers some sort of psychological disorder and infuses fearfulness in one's mind. Absolute silence on the living planet, however, is almost impossible. A place with absolute silence can be created only by making it sound-proof with the aid of some material.

Mild sounds of about 10 dB are necessary for the normal health and well-being of humans. Experiences of a variety of sounds are inevitable. Dynamic nature creates various sounds many of which, for example, the chirping of birds, are so pleasant to mind! Sounds of nature enhance one's happiness and are necessary for cultivating creativity in human minds. Music and songs with soft sounds contribute to striking a balance of mind and elevating the level of happiness and creativity. Hearing is one of the five senses human beings have integrated within the system. Without exposure to a sound, which, of course, is unlikely, this sense will go defunct and human development processes will be impaired. Therefore, living amidst and working amidst mild sounds, or keeping away from a sound-polluted environment is vitally essential for human beings for their balanced development and sound health.

Effect on Hearing Ability

Impairment of hearing ability is to be the first to be affected upon prolonged exposure to sound pollution. Sudden loud noise and explosive sounds (above 110 dB) may lead to acoustic trauma and permanently damage the eardrum. A temporary hearing loss is also referred to as a threshold shift.

An explosion of 140 dB and one's exposure to this magnitude of sound may even cause permanent loss of hearing. Frequent sounds of very high intensity may lead to permanent damage to sensory cells responsible for hearing. People residing in noise-ridden areas, such as near railway stations, and highways, or working in noisy factories often suffer from deafness.

Recent studies suggest that continuous long-term exposure to 55 dB can also cause permanent hearing damage (Ahluwalia and Malhotra 2018).

Effect on General Health

People compelled to live in noisy areas may become victims of physical, physiological, psychological, and emotional ailments. Some of the effects of noise pollution on general human health are given below:

- Noise causes stress, anxiety, and mood upset. It loses one's concentration and reduces work output. A sound becomes painful at 120 dB.
- Detraction of attention, insomnia, and emotional breakdown may be caused due to noise pollution. Extreme cases may even cause fright.
- Dilation of the eye pupil, impairment of night vision, reduced color perception, and digestive spasms may also be caused by noise pollution.
- Continuous exposure to noise pollution causes hormonal imbalances leading to several health disorders, such as increased heartbeat, constriction of blood vessels, increased cholesterol level, high blood pressure, decreased heart output, and hypertension.
- Explosive sounds (above 140 dB), such as those generated by crackers, motorcycles, trains, etc. create a sudden and startling reaction that may cause disturbance in psychomotor performance and even heart failure.
- People living in noisy areas may also suffer from psychiatric illnesses.
- A sonic boom produces a startling effect that may even lead to brain damage.
- Noise pollution equally affects animals. Liver, heart, and brain damage have been reported in animals. Some animals, notably rabbits, die of explosive sounds that crackers or dynamite blasts generate.

On many occasions, sound pollution is taken in good spirits. It all depends on the mood of the people and is also occasion-specific. The use of DJs is customarily made and enjoyed on several occasions even though noise created by this instrument is deleterious to health. The 21-Cannon Salute on special occasions is celebrated on special occasions. Such sound-dependent incidences are not regarded as pollution and are rather enjoyed and celebrated.

Control of Sound Pollution

There can be two ways to avoid sound pollution: i) do not generate sound, and ii) reduce the sound intensity to a minimum so that it is not felt as noise. The first solution applies only to human-generated sounds. Natural sounds are seldom perceived as pollutants. Human-generated sound pollution is difficult to be avoided in our contemporary world of sounds. If the sound generation is not minimized to a point where it is not felt as pollution, we can only reduce the impact of sound pollution which can be done in two ways: i) by changing or modifying the path of sound transmission, and ii) by preventing or minimizing sound intensity reaching the receiver.

Some of the ways of preventing and/ or minimizing the impact of sound pollution are as follows:

- The noise of motor vehicles can be reduced by planting many rows of trees on roadsides. Plants not only purify the air, regulate the water cycle and maintain climatic balance, but also absorb sound pollution to a considerable level.
- The plantation of trees and shrubs in industrial areas, residential areas, and commercial establishments will greatly contribute to reducing sound pollution. Vegetation reduces noise pollution through the absorption, refraction, and deflection of sound waves. It has been shown that in some situations, trees can reduce noise levels by up to 50%. Beautification of the sights and other attributes of trees will also be accrued to people. The establishment of parks with trees, shrubs, and herbs within cities and towns will significantly contribute to decreasing sound pollution. Vegetation is perhaps the best natural barrier against sound pollution. In addition to reducing the impact of artificial sounds, vegetation also provides an environment of natural pleasant sounds, like birds' songs, musical sounds generated by air passing through tree canopies, and the likes.
- Noise-producing industries, factories, aerodromes, railway stations, bus stands, etc. should be located far away from residential areas. This would help people get rid of continuous exposure to sound pollution.
- Silence zones must be created around educational institutes, hospitals, and residential areas. This would greatly reduce the effects of noise pollution on people leading to an increase in their working efficiency and creativity.
- Machines and equipment that produce noise pollution must be replaced with those generating sounds of low intensity. There are technologies available to manufacture such machines and equipment. If that is not workable in some cases, then noisy machines should be installed in sound-proof chambers. Proper maintenance of the machines, such as the use of lubricants, would also help reduce noise pollution.
- Workers on duty in noisy industries or noisy places have to suffer more from noise pollution than others. Such occupational exposure to sound

pollution can be decreased by using noise-protective devices such as ear muffs for protection against sounds above 90 dB and cotton plugs for protection against sounds below 90 dB.

- Certain rules and regulations have already been framed by central and state pollution control boards in India. Legal measures should be strictly implemented and followed to control pollution generated in social-cultural ceremonies, festivals, public speeches, etc.

The Central Pollution Control Board (CPCB) has recommended zone-wise permissible ambient noise levels as presented in Table 5.2. It depends on time length according to human activities. During day time, for example, a residential area can remain unaffected by a sound level of up to 55 dB but at night when people are to rest and rejuvenate themselves, the sound level must not exceed 45 dB.

Table 5.2: Zone-wise permissible ambient noise levels

Zone	Sound level (dB)	
	Day time (06:00 to 21:00 hours)	Night time (21:00 to 06:00 hours)
Silent	50	40
Residential	55	45
Commercial	65	55
Industrial	75	70

The world is awakening against noise pollution and all countries have their legal provisions to prevent and control noise pollution. The World Health Organization (WHO) has also set up necessary standards to minimize and control noise pollution.

Summary

Sound pollution, or noise pollution, is undesirable and unwanted sound that becomes an irritant and a source of stress and often of some abnormalities and diseases. Some technological developments cannot work without creating sound in the amounts that can be dangerous to life, such as those associated with the transport sector, industrial plants/ factories, construction machinery, defense machinery, domestic gadgets, entertainment sources, public address system, etc.

Frequency and intensity are the two very important characteristics of a sound. Sound measurement has been devised on a logarithmic, rather than linear, scale with a reference to 10 and referred to as decibel. The range of the noise level can extend from zero to 140 dB. Beyond 120 dB, the noise causes physical discomfort. Normal conversation sound ranges between 35 dB and 60 dB.

Impairment of hearing may be caused if one is exposed to 80 dB or more for a longer period.

Prolonged exposure to sound intensity of 120 dB is painful. Continuous exposure to the noisy environment in some factories can make the workers lose their ability to hear soft sounds. Sudden loud noise and explosive sounds (above 110 dB) may lead to acoustic trauma and permanently damage the eardrum. An explosion of 140 dB and one's exposure to this magnitude of sound may even cause permanent loss of hearing. Frequent sounds of very high intensity may lead to permanent damage to sensory cells responsible for hearing. Noise causes stress, anxiety, mood upset, detraction of attention, insomnia and emotional breakdown, dilation of the eye pupil, impairment of night vision, reduced color perception, digestive spasms, etc. Continuous exposure to noise pollution causes hormonal imbalances leading to several health disorders, such as increased heartbeat, constriction of blood vessels, increased cholesterol level, high blood pressure, decreased heart output, and hypertension.

There can be two ways to avoid sound pollution: do not generate sound, and reduce the sound intensity to a minimum so that it is not felt as noise. Proper maintenance of the machines, such as the use of lubricants, would also help reduce noise pollution. The noise of motor vehicles can be reduced by planting many rows of trees on roadsides. The plantation of trees and shrubs in industrial areas, residential areas, and commercial establishments will greatly contribute to reducing sound pollution. Noise-producing industries, factories, aerodromes, railway stations, bus stands, etc. should be located far away from residential areas. Silence zones must be created around educational institutes, hospitals, and residential areas.

Exercises

A. Multiple Choice Questions

1. The sound travels fastest through

   a. Solid　　b. Liquid

   c. Air　　d. Space

2. The frequency of sound is a measure of

   a. The loudness of the sound
   b. How low or how high a sound is
   c. How sound affects the human mind
   d. The sound propagates through a liquid

3. The intensity of sound is a measure of the
   a. The energy delivered by sound
   b. Wavelength of sound
   c. Source of sound
   d. Impact of sound
4. The frequency and intensity of a sound are also called
   a. Pitch and loudness, respectively
   b. Loudness and pitch, respectively
   c. pain and damage, respectively
   d. Vibrations and threshold, respectively
5. "Decibel" is chosen as a unit to measure sound to give honor to
   a. Richard Campbell
   b. Alexander Graham Bell
   c. John Dumbell
   d. Robin Stumbell
6. The human ear cannot detect
   a. ultrasound
   b. infrasonic sound
   c. supersonic sound
   d. the sound created by jet aircraft at an altitude of 8000 m
7. The decibel scale is
   a. Arithmetic
   b. Algebraic
   c. Logarithmic
   d. Geometric
8. Whispering causes a sound range of
   a. 10 dB
   b. 15 dB
   c. 30 dB
   d. 120 dB
9. Hearing loss due to sound pollution depends on
   a. Intensity of sound
   b. Duration of sound
   c. Intensity and duration of the sound
   d. None of the above

10. An explosion of 140 dB and one's exposure to this magnitude of sound can cause

    a. Blindness
    b. Baldness
    c. Permanent deafness
    d. Lameness

11. What level of noise, as recent studies suggest, can cause in a person who has continuous long-term exposure to it?

    a. 10 dB
    b. 20 dB
    c. 30 dB
    d. 55 dB

12. What level of noise is dangerous for a person with prolonged exposure to it?

    a. 0 dB
    b. 55 dB
    c. 80 dB
    d. All of the above

13. Which of the following sources does not generate sound pollution?

    a. Domestic appliances
    b. Construction Equipment
    c. Library
    d. Motor vehicles

14. How can noise pollution deteriorate your health?

    a. It may not let you have proper sleep
    b. It may give you a headache and damage your hearing power
    c. It may make you irritate
    d. In all the ways mentioned above

15. Which body in India has recommended zone-wise permissible ambient noise levels?

    a. University Grants Commission (UGC)
    b. Indian Council of Agricultural Research (ICAR)
    c. Council of Scientific and Industrial Research (CSIR)
    d. Central Pollution Control Board (CPCB)

16. What level does a sound become painful at?

    a. 120 dB
    b. 110 dB
    c. 100 dB
    d. 80 dB

17. Temporary hearing loss is also referred to as
    a. Temporary hearing shift
    b. Temporary threshold shift
    c. Temporary ear pain
    d. Temporary deafness
18. The ambient noise level of a residential area during day and night time, respectively, is
    a. 30 dB and 40 dB
    b. 55 dB and 45 dB
    c. 45 dB and 55 dB
    d. 65 dB and 60 dB
19. What is recommended for the control of noise pollution caused by transport vehicles?
    a. Plantation of trees in multiple layers on roadsides
    b. Increased use of diesel, rather than petrol, vehicles
    c. Minimum movement of trucks
    d. Increased use of bicycles on highways
20. A sound measures 120 dB at a point. At 100 meters away from the point, it would measure
    a. More than 120 dB
    b. Less than 120 dB
    c. Equal to 120 dB
    d. Depending upon the source of the sound

Answers: 1-**a**, 2-**b**, 3-**a**, 4-**a**, 5-**b**, 6-**b**, 7-**c**, 8-**c**, 9-**c**, 10-**c**, 11-**d**, 12-**d**, 13-**c**, 14-**d**, 15-**d**, 16-**a**, 17-**b**, 18-**b**, 19-**a**, 20-**b**

B. Short-Answer Questions

1. How would you define a sound?
2. Mention the natural sources of the sound.
3. Define the frequency of a sound.
4. What is the intensity of a sound?
5. What is the decibel scale?
6. Which industries are the major causes of noise pollution?
7. Give some examples of noise pollution caused during social ceremonies.

8. Mention some health hazards caused by noise pollution.
9. Which places are often highly noisy?
10. How can the sound pollution caused by vehicles and traffic be reduced?

C. Long-Answer Questions

1. Discuss the various sources of sound pollution giving striking examples.
2. Our world is increasingly becoming a victim of noise pollution. Discuss your perspective.
3. How is sound pollution affecting human societies? Make a discussion focused on your own experiences of the noise-ridden human habitats.
4. Discuss various measures necessary to control sound pollution.
5. Write short notes on the following:
   (a) Measurement of sound
   (b) Psychological effects of noise pollution
   (c) Legal provisions relating to noise pollution

Suggested Readings

Ahluwalia, V. K. and Malhotra, S. 2018. *Environmental Science*. New Delhi: Ane Books Pvt. Ltd. 358 pp.

Singh, Vir. 2014. Biosphere vs technosphere. *The Speaking Tree – The Times of India.* April 20, 2014.

Web Resources

http://www.quiet.org/faq.htm
https://education.nationalgeographic.org/resource/noise-pollution
https://en.wikipedia.org/wiki/Noise_pollution
https://moef.gov.in/en/rules-and-regulations/environment-protection/noise-pollution/
https://www.speakingtree.in/article/biosphere-vs-technosphere
https://www.vedantu.com/biology/causes-of-noise-pollution

6

Radioactive Pollution

Intelligentsia/ Pexels

"For fifty years, nuclear power stations have produced three products which only a lunatic could want: bomb-explosive plutonium, lethal radioactive waste and electricity so dear it has to be heavily subsidised. They leave to future generations the task, and most of the cost, of making safe sites that have been polluted half-way to eternity." – James Buchan

Radioactive materials are actual and potential pollutants. Spreading through air, water, and soil, the radioactive pollutants leave their extremely deleterious effects on organisms. Among all types of pollutants, radioactive pollutants are the most deadly ones. The nature of contamination and impacts of radioactive pollution are also different from other types of pollutants and pollution. Although living organisms are often exposed to a certain degree of radiation pollution, radiation emanating from radioactive substances has increased in recent decades and is a special and serious concern.

Types of Radiation and Their Characteristics

Radioactive pollution is largely attributable to radioactivity, a natural property of certain elements, such as radium, uranium, thorium, etc. of spontaneously emitting protons (alpha particles), electrons (beta particles), and gamma rays (short-wave electromagnetic waves) through the disintegration of their atomic nuclei (nuclides).

An alpha particle consists of two protons and two neutrons, which is akin to a helium nucleus. Alpha particles travel at a very slow speed, whereas gamma rays propagate at the speed of light. The alpha particles have less penetration power than the beta and gamma rays. In human tissues, alpha particles can penetrate up to a depth of about 0.005 to 0.008 cm. The gamma rays, on the other hand, are the most penetrating and can easily pass through a human body without striking anything in between. The high penetration power of gamma rays is because they have no charge; therefore, they do not interact with anything as the charged alpha particles would do. The ability of radiation to damage molecules is analyzed in terms of its ionization power. The ionizing power of the alpha particles is very high, while of gamma rays it is almost nil.

Beta particles are much smaller and have much less ionizing power and much greater penetration power than alpha particles. Characteristics of beta particles, in other words, lie between those of alpha particles and gamma rays. A comparison between the three has been made in Table 6.1.

Table 6.1: Characteristics of three types of radiation

Particle	Mass	Penetrating power	Ionizing power
Alpha (α)	4	Very low	Very high
Beta (β)	1/1840	Moderate	Moderate
Gamma (γ)	0	Very high	Very low

Radioisotopes are the radioactive isotopes of an element with the same number of protons as their atomic nuclei but differing numbers of neutrons, which spontaneously undergo radioactive decay by emission of radiation. In other words, radioisotopes are atoms with an unstable combination of protons and neutrons. Each isotope has its characteristic emissions.

An important concept related to radioisotopes is their half-life. Radioactive half-life is the time required for the radioisotopes to decay by half. Some of the radioisotopes have a very short half-life. Examples are: Po^{215} has a half-life of 0.0018 seconds, Bi^{212} of 60.5 seconds, and Po^{218} has just 3 minutes. C^{14} has a half-life of 5570 years which is an intermediate value. Some radioisotopes have a very high value of half-life. Examples are: U^{235} has a half-life of 7000 million years and U^{238} of 4.5 billion years.

Radioactivity is expressed by various units, *Curie* (Ci) being a commonly used one. Ci unit is named after Polish chemist Marie Curie, who, along with her husband Pierre Curie, discovered Radium and Polonium. The latter was named after Poland, the country Marie Curie belonged to. A Curie is the amount of radioactivity from 1 g of Radium-226 that undergoes about 37 billion nuclear transformations per second (Ahluwalia and Malhotra 2018).

The International System (SI) measures radioactive decay in *Becquerel* (Bq). One *Becquerel* is the activity of a quantity of a radioactive material in which one nucleus decays. Therefore, 1 Ci equals 3.7×10^{10} decays per second. In the case of environmental studies, the actual dose of radiation is generally considered and it is measured in *rads* (rd) and *rems*. In International System (SI), the absorbed dose of radiation is *grays* (Gy) and *sieverts* (Sv).

1 Gy = 100 rads; 1 Sv = 100 rem

Sources of Radioactive Pollution

Natural Sources

The emission of radiation is a phenomenon sustained since the birth of the cosmos. Cosmic rays coming from the sun and space are natural radiation. The intensity of this radiation reaching the Earth depends on altitude and some other environmental factors. Living organisms are exposed to cosmic rays reaching the Earth and terrestrial radiation emanating from radionuclides present in the earth's crust, for example, C^{14}, U^{238}, Ra^{224}, Th^{232}, and K^{40}. The cosmic and terrestrial radiation together is called background radiation.

Anthropogenic Sources

Anthropogenic or man-made sources include mining and refining of radioactive substances like thorium (Th) and plutonium (Pu), preparation of radioactive isotopes, laboratory experiments and research on radioactive isotopes, radiation therapy, nuclear weapon tests, and nuclear fuels and nuclear power plants.

Radiation Therapy

Human exposure to high-energy x-rays is quite common. Radiation therapy is also given in cancer treatment. While such radiation therapy can cure some diseases, overdoses of the radiation might have adverse effects on the body.

Atomic Explosion

This world of ours is seated on a stockpile of nuclear weapons. All nuclear weapons have come into being through nuclear tests. The radiation is produced during the process of fission as well as fusion. Nuclear weapons use uranium-235 (U^{235}) and plutonium-239 (Pu^{239}) for fission and hydrogen or lithium as fusion material. Nuclear fallout is what occurs as a result of atomic explosions. The radioactive particles are thrown up into the sky like clouds and from there they gradually settle on vegetation, inhabited areas, water, and soil. Nuclear tests and the use of atomic or hydrogen atoms, thus, end up with widespread contamination of earth's ecosystems. The nuclear fallout contains highly dangerous and life-annihilating elements such as strontium-90 (Sr^{90}), cesium-137 (Ce^{137}), and iodine-131 (I^{131}).

In earth's ecosystems, radioactive elements do not just settle down in one place. From the soils, they enter into plants and subsequently into herbivores and human beings. Radioactive elements enter into carnivores through food chains and that too in increasing concentrations at every successive trophic level. From the soils, radioactive pollutants are washed into aquatic ecosystems through rainwater. Not only terrestrial food chains, but radioactive substances also happen to flow along aquatic food chains. From fish and other sea organisms, these elements enter into human systems. Not only terrestrial and aquatic food chains, but the elements of radioactive fallout also enter into detritus food chains. When detritus-feeding organisms die and decay, these elements release into the soil and resume the same pathways.

Nuclear Fuels and Power Plants

Radioactive isotopes are used as fuel to trigger the fission process in nuclear reactors (see Fig. 6.1). In this process, a huge amount of energy is generated in the form of heat which is used for producing steam to turn large turbines to generate electricity. The process is not as simple as it seems to be. The radioactive elements and the coolants used in the process phenomenally contribute to radiation pollution. The most tedious problem is waste disposal. In the process of electricity generation, nuclear reactors release highly dangerous wastes in the form of solids, liquids, and gas. Although nuclear reactors are supposed to be made leak-proof, that cannot be taken guaranteed forever. Several incidents of leakages and widespread radioactive pollution have been witnessed in many cases, notably in the Three Mile Island nuclear power plant in the USA in 1979, Chornobyl (Chernobyl) nuclear power plant in the erstwhile USSR in 1986, and Fukushima nuclear power plant in Japan in 2011.

Fig. 6.1: Radioactive pollution is the worst kind of pollution that the biosphere faces and experiences (image by Johanne Plenio on Pexels)

Escaping radioactive wastes into the water, air, and soil can bring in a trail of grave consequences. Liquid effluents carrying radioisotopes generated in the process of electricity generation in an atomic reactor may enter into water streams and subsequently into the food chain. Halogen and inert gases escape into the atmosphere as vapors, come down to land and surface waters with rainwater and cause havoc to life.

Waste disposal containing radionuclides is a very complex process. Moreover, this is a very expensive process. Their storage is also quite problematic. Conventionally, the spent fuel (burnt-up nuclear fuel) is disposed into burial grounds or some containers after its reprocessing. Carrying the spent fuel to reprocessing plant, which is often located at some distance from the nuclear reactor, might be quite risky. An accidental escape of the spent fuel may contaminate a very large area with dangerous radionuclides.

Research Experiments

The use of radioisotopes in research experiments has become a very common phenomenon in science. Often used isotopes in laboratories are ^{14}C, ^{32}P, and ^{125}I. There are chances that these isotopes, even in traces, pass into wastewater from laboratories and mix with natural water bodies.

Effects of Radioactive Pollution

The effects of a radioactive substance would vary according to its half-life, radiation intensity, rate of diffusion, and concentration or rate of deposition (Sharma 2009). Some physical or abiotic environmental factors, like rainfall, winds, atmospheric temperature, etc. would also influence the effects of radioactive pollutants.

Based on their mode of action on living cells, the radiations can be classified as non-ionizing and ionizing radiations.

Non-ionizing Radiations

They are short-wave radiations as part of solar radiations, such as ultraviolet rays. These radiations have low penetration power and are harmful to multicellular organisms when they are exposed to such radiation for a longer period.

- Long exposure to non-ionizing radiation can damage the eyes. Light reflection from snow may cause snow blindness. Eye damage may also occur due to exposure to light reflection from coastal sands. Looking at the eclipsing sun is often advertised as eye-damaging.
- These radiations cause injury to skin cells and blood capillaries and cause what is called sunburn.
- Their long-term effect may lead to the inactivation of DNA and RNA and the incidence of mutations and cancer.

Ionizing Radiations

Ionizing radiations have a high degree of penetrability and physically break down biomolecules and damaged cells, a property called ionization. X-rays, radiations from radioactive elements or atomic radiations, and cosmic rays are examples of ionizing radiations.

The molecular effects of the ionizing radiations may be short-range or immediate and long-term or delayed effects.

1. Short-range effects are often acute and are revealed within a few days or weeks after the organisms are exposed to radiation. Skin burns, physical crippling, impaired metabolism, dead tissues, and death of the exposed organisms are the indicators of the short-range effects of ionizing radiations.
2. Long-range molecular effects take a longer time to express. The delayed effects owing to ionizing radiations especially make scientists interested to carry out research projects. The long-range effects include genetic changes, loss of vitality, anemia, point mutations, chromosomal aberrations, increased chances of tumor and cancer, hemorrhages, and a shortening of life span.
3. The mutated genes may persist in the radiation-affected organisms and may affect their progeny.

Acute doses of radiation affect all organisms. Among human beings, the especially sensitive organs are the thyroid gland, spleen, lymph nodes, and bone marrow. Radiation-emitting isotope ^{131}I readily enters into the food chain and, as human beings are the major consumers, its magnified concentrations affect in various ways by damaging white blood cells, lymph nodes, spleen, and bone marrow and may also cause defective eyesight, sterility, and lung and skin cancer.

Accumulation of some radioactive elements in plant and animal bodies is species-specific. For example, marine animals selectively deposit ^{90}Sr, fish bodies accumulate ^{55}Fe, and oysters ^{65}Zn. ^{90}Sr can replace Ca in plants and animals. In bones, it causes tissue degeneration and bone cancer.

The biological effects of the ionizing radiations may also be divided into somatic and genetic effects. The short-range somatic effects include radiation sickness and acute radiation syndrome, whereas long-range somatic effects include leukemia, cancer, fetal abnormalities, and a shortening of life. The genetic effects of ionizing radiations include chromosomal aberrations and point mutations or mutations.

Protection from and Control of Radioactive Pollution

There is no cure once an organism is affected by radiation. Protection from harmful radiation is the best measure.

- Wearing dark glasses can prevent the eyes from the effect of ultraviolet rays. Visible light also diminishes the effects of ultraviolet rays considerably.

- Appropriate measures should be strictly adopted to prevent occupational exposure.
- Nuclear reactors must be leak-proof. It must be ensured that no radioactive element leaks during the processing of radioisotopes.
- No leakage of radioactive elements should be there during experimentation in laboratories.
- Safe disposal of radioactive wastes must be strictly followed. They must be converted into harmless form and stored at a leak-proof place or in containers where they keep decaying harmlessly.
- Regular monitoring of the sinking place or dumping areas must be ensured.

Nuclear fallout is difficult to control completely. However, its hazards may be minimized by taking up certain precautions.

- High-level wastes with high radioactivity per unit volume must be contained in underground tanks without treatment. Liquids should be converted into solids, such as ceramics, and then buried deep underground following safety precautions.
- Low-level wastes with very low radioactivity must be dispersed into the environment in such a way that they are not allowed to become part of food chains.
- In this world seated on a huge stockpile of nuclear arms, further nuclear testing and use of nuclear weapons must be banned through evolving and implementing pertinent international laws.

There are several organizations, notably the International Atomic Energy Agency (IAEA) and Ottawa-based international non-government organization, the International Commission on Radiological Protection (ICRP) that recommend and disseminate safety-related standards.

Summary

Among all types of pollutants, radioactive pollutants are the most deadly ones. Radioactive pollution is largely attributable to radioactivity, a natural property of certain elements, such as radium, uranium, thorium, etc. of spontaneously emitting protons (alpha particles), electrons (beta particles), and gamma rays (short-wave electromagnetic waves) through the disintegration of their atomic nuclei (nuclides). The ionizing power of the alpha particles is very high, while of gamma rays it is almost nil. Beta particles are much smaller and have much less ionizing power and much greater penetration power than alpha particles. Radioisotopes are the radioactive isotopes of an element with the same number of protons in their atomic nuclei but differing numbers of neutrons, which spontaneously undergo radioactive decay by the emission of radiation. Each isotope has its characteristic emissions. Radioactive half-life is the time required for many radioisotopes to decay by half.

Radioactivity is expressed by various units, *Curie* (Ci) being a commonly used one. A Curie is the amount of radioactivity from 1 g of Radium-226 that undergoes about 37 billion nuclear transformations per second. The International System (SI) measures radioactive decay in *Becquerel* (Bq). One *Becquerel* is the activity of a quantity of a radioactive material in which one nucleus decays. Therefore, 1 Ci equals 3.7×10^{10} decays per second. In International System (SI), the absorbed dose of radiation is *grays* (Gy) and *sieverts* (Sv). 1 Gy = 100 rads; 1 Sv = 100 rem

Cosmic rays coming from the sun and space are natural radiation. Living organisms are exposed to cosmic rays reaching the Earth and terrestrial radiation emanating from radionuclides present in the Earth's crust, for example, C^{14}, U^{238}, Ra^{224}, Th^{232}, and K^{40}. The cosmic and terrestrial radiation together is called background radiation. Anthropogenic or man-made sources include mining and refining of radioactive substances like thorium (Th) and plutonium (Pu), preparation of radioactive isotopes, laboratory experiments and research on radioactive isotopes, radiation therapy, nuclear weapon tests, and nuclear fuels and nuclear power plants.

From the soils, the radioactive substances enter plants and subsequently into herbivores and human beings. Radioactive elements enter into carnivores through food chains and that too in increasing concentrations at every successive trophic level. From the soils, radioactive pollutants are washed into aquatic ecosystems through rainwater.

Non-ionizing radiations cause injury to skin cells and blood capillaries and cause what is called sunburn. Their long-term effect may lead to the inactivation of DNA and RNA and the incidence of mutations and cancer. Ionizing radiations have a high degree of penetrability and physically break down biomolecules and damage cells, a property called ionization. The long-range effects include genetic changes, loss of vitality, anemia, point mutations, chromosomal aberrations, increased chances of tumor and cancer, hemorrhages, and a shortening of life span. The mutated genes may persist in the radiation-affected organisms and may affect their progeny. Acute doses of radiation affect all organisms.

There is no cure once an organism is affected by radiation. Protection from harmful radiation is the best measure. Wearing dark glasses can prevent the eyes from the effect of ultraviolet rays. Appropriate measures should be strictly adopted to prevent occupational exposure. Nuclear reactors must be leak-proof. Safe disposal of radioactive wastes must be strictly followed. Regular monitoring of the sink or dumping areas must be ensured. Nuclear fallout is difficult to control completely. However, its hazards may be minimized by taking up proper precautions.

Exercises

A. Multiple Choice Questions

1. What source of radiation travels at a very slow speed?
   a. Alpha particles
   b. Beta particles
   c. Gamma rays
   d. All propagate with equal speed
2. Which of the following radiation has no charge?
   a. Alpha particles
   b. Beta particles
   c. Gamma rays
   d. All of the above
3. Cosmic radiation and terrestrial radiation together are known as
   a. Solar radiation
   b. background radiation
   c. Planetary radiation
   d. Radioactive radiations
4. An alpha particle consists of
   a. Two protons
   b. Two neutrons
   c. Two protons and two neutrons
   d. Two protons and two electrons
5. An alpha particle, in other words, is a
   a. hydrogen nucleus
   b. polonium nucleus
   c. uranium nucleus
   d. helium nucleus
6. Which of the following has the least penetrating power?
   a. Alpha particles
   b. Beta particles
   c. Gamma particles
   d. X-rays

7. In India, the uranium mines are located in the State of

   a. Uttarakhand
   b. Jharkhand
   c. Assam
   d. Haryana

8. In India, the task of surveying and exploration of uranium, thorium, rare metals, and rare-earths is assigned to

   a. Atomic Minerals Directorate for Exploration and Research (AMD)
   b. Uranium Corporation of India Ltd. (UCIL)
   c. Indian Rare Earths Ltd. (IREL)
   d. Council of Scientific and Industrial Research (CSIR)

9. Atomic explosions and nuclear tests result in the contamination of ecosystems by highly dangerous elements like strontium-90 (Sr^{90}), cesium-137 (Ce^{137}), and iodine-131 (I^{131}). This process is known as

   a. Nuclear wastage
   b. Nuclear leakage
   c. Nuclear fallout
   d. Nuclear hazards

10. Which of the following has the shortest half-life, just about 3 minutes?

    a. Po^{218}
    b. C^{14}
    c. Cs^{137}
    d. U^{235}

11. Who did discover Polonium and named it after his/ her/ their country, Poland?

    a. Becquerel
    b. Marie Curie and Pierre Curie
    c. Lavoisier
    d. Albert Einstein

12. An accident at a nuclear power station took place at the Three Mile Island in 1979. What country is the Three Mile Island located in?

    a. Great Britain
    b. Russia
    c. France
    d. USA

13. Chornobyl Nuclear Reactor accident occurred in

    a. 1986
    b. 1996
    c. 2001
    d. 2011

14. What ailment is caused due to prolonged exposure to high-intensity radiation?

    a. Diabetes
    b. Mutation
    c. Urological ailments
    d. Gastric problems

15. Which of the following radioactive elements may cause thyroid cancer?

    a. U-235
    b. Cs-127
    c. I-131
    d. Po-128

16. One of the worst nuclear accidents took place in 2011 at

    a. Chernobyl
    b. Three Mile Island
    c. Fukushima
    d. none of the above

17. The effects of nuclear radiation include

    a. Somatic effects
    b. Genetic effects
    c. somatic as well as genetic effects
    d. neither somatic nor genetic effects

18. In plants and animals, ^{90}Sr can replace

    a. Ca
    b. Fe
    c. S
    d. K

19. Among human beings, the organs especially sensitive to harmful radiation are

    a. Thyroid gland and spleen
    b. Lymph nodes
    c. Bone marrow
    d. All of the above

20. The headquarters of the International Commission on Radiological Protection (ICRP) are located in

    a. Geneva, Switzerland
    b. Ottawa, Canada
    c. Berlin, Germany
    d. New York, USA

Answers: 1-**a**, 2-**c**, 3-**b**, 4-**c**, 5-**d**, 6-**a**, 7-**b**, 8-**a**, 9-**c**, 10-**a**, 11-**b**, 12-**d**, 13-**a**, 14-**b**, 15-**c**, 16-**c**, 17-**c**, 18-**a**, 19-**d**, 20-**b**

B. Short-Answer Questions

1. What do you mean by radioactive pollution?
2. Mention the natural sources of radioactive pollution.
3. What is non-ionizing radiation?
4. Mention in brief the characteristics of alpha, beta, and gamma particles.
5. What is nuclear fallout?
6. How is radioactive pollution caused by research experiments?
7. Define radioisotopes.
8. What is a nuclear reactor?
9. Name some accidents that have occurred at nuclear power plants.
10. Mention some ailments caused due to human exposure to harmful radiation.

C. Long-Answer Questions

1. What are various sources of radioactive pollution? Describe with examples.
2. Describe various effects of radioactive pollution citing some examples.
3. What are various ways of protecting against and control of radioactive pollution? Elaborate with perspectives on safety, technology, and policy.
4. Effectively and consciously used for environmental conservation and socioeconomic development, nuclear energy could be a boon for the world. What are your views in this regard? Describe especially in the context of the ongoing climate change and its negative consequences on socio-economic development.
5. Write short notes on the following:

   (a) Nuclear fuels and power plants

   (b) Non-ionizing and ionizing radiation

   (c) Nuclear accidents

Suggested Readings

Ahluwalia, V. K. and Malhotra, S. 2018. *Environmental Science*. New Delhi: Ane Books Pvt. Ltd. 368 pp.

Sharma, J. P. 2013. *Environmental Studies*. Bangalore: University Science Press. 221 pp.

Web Resources

https://en.wikipedia.org/wiki/Radioactive_contamination

https://nuclearsafety.gc.ca/eng/resources/radiation/introduction-to-radiation/radiation-health-effects.cfm

https://onlinelibrary.wiley.com/doi/10.1002/9781118914236.ch36

https://www.conserve-energy-future.com/radioactive-pollution-causes-effects-solutions.php

https://www.environmentalpollutioncenters.org/radiation/

7

Global Warming and Climate Change

Augustin Lautaro/ Unsplash

"One thing leads to the other. Deforestation leads to climate change, which leads to ecosystem losses, which negatively impacts our livelihoods – it's a vicious cycle."

– Gisele Bundchen

Global warming is the hottest issue facing life in our times. The gradual average increase in air temperatures near Earth's surface over the past one to two centuries is leading to the phenomenon called global warming. It is this global phenomenon that triggers climate change, threatening the entire life on Earth. Earth's climate has changed over all conceivable geological timescales, but that was all on account of the natural phenomenon. But, human activities are making the phenomenon more pronounced since the beginning of the Industrial Age, now emerging as a dominant triggering factor in influencing the overall climate fabric of the planet.

Greenhouse Gases

The role of greenhouse gases in the atmosphere is to hold Earth's surface temperature within a range appropriate for the physiological well-being and functioning of living organisms. Global warming is a consequence of greenhouse gases accumulating disproportionately in the atmosphere and attaining a level beyond a limit. The proportion of these gases in the atmosphere determines the average global temperature.

A greenhouse gas (GHG) has the characteristic of absorbing and emitting radiation within the thermal infrared range. The primary GHGs in the Earth's atmosphere is carbon dioxide (CO_2), methane (CH_4), nitrous oxide (N_2O), ozone (O_3), and water vapor (H_2O). The secondary GHGs include chlorofluorocarbons (CFCs) and hydrofluorocarbons (HFCs).

Nitrogen (N_2), oxygen (O_2), and argon (Ar) constitute 78%, 21%, and 0.9% proportion of the atmosphere respectively, thus making up 99.9% of the atmospheric composition. These are all non-GHGs. Most of the atmosphere, in essence, consists of the non-GHGs and less than 0.1% of gases include the GHGs, which, in order of their abundance are as follows: $H_2O > CO_2 > CH_4 > N_2O >$ CFCs > HFCs.

The other GHGs, such as Ozone (O_3), carbon monoxide (CO), and nitrogen oxides (NOx) are short-lived and generally not counted among the main GHGs. Aerosols (such as black carbon and mineral dust) that would vary over different locations and times are also not generally counted towards GHGs.

Although water vapor is a potential GHG contributing to the direct greenhouse effect to the tune of 36 to 72%, it is generally not a matter of concern when we discuss the fate of other GHGs. First, atmospheric water vapor concentration (10–50,000 ppm) fluctuates from region to region and area to area in the same region, as well as according to seasons, day to day in the same season, and even from time to time in the same day. Second, the water cycle is vital for the ecological integrity of the biosphere and its increased concentration in the atmosphere is generally of positive thermodynamic bearing.

Sources of the GHG emissions and various characteristics of the main GHGs have been presented in Table 7.1.

Table 7.1: The main Greenhouse gases, their sources and characteristics

Greenhouse gas	Source of emission	Concentration in 1750	Concentration in 2011*/ 2019**	% increase since 1750	Atmospheric life time (years)
Carbon dioxide (CO_2)	• Chiefly produced by the burning of fossil fuels; • Released by plants and animals during respiration	280 ppm	415 ppm**	48	5-200
Methane (CH_4)	• Caused by anaerobic decomposition in the presence of methanogens, a group of bacteria; • Released from garbage dumps, swamps, flooded rice fields and enteric fermentation	700 ppb	1801 ppb*	157	12
Nitrous oxide (N_2O)	• Industry, biomass burning and agriculture are the main sources; • Breakdown of nitrogenous fertilizers in the soil, burning of N-rich fuels, livestock wastes and nylon production	270 ppb	324 ppb*	20	114
Chlorofluorocarbons (CFC-11) + Hydrofluorocarbons (HFC-23)	• Synthesized gaseous compounds of carbon and halogens; • Leaking of air conditioners and refrigeration units, propellants in aerosol spray cans, production of plastic foams, and evaporation of industrial solvents, • Extensively utilized as refrigerants, aerosol propellants, fire extinguishers and insulators	0	262 ppt*	262	45 – 260

ppm = parts per million (10^6) = µmol/mol; ppb = parts per billion (10^9) = nmol/mol; ppt = parts per trillion (10^{12}) = pmol/mol

In addition to the major GHGs listed above, the other gases that are given in the IPCC list (IPCC 2001) include perfluorocarbons, sulfur hexafluoride (SF_6), tetrafluoromethane (CF_4), hexafluoroethane (C_2F_6), nitrogen trifluoride (NF_3). However, as these gases are in extremely low proportion, they are often omitted from the main list of greenhouse gases.

Greenhouse Effect

Among the non-greenhouse gases that make up 99.9% of the atmosphere, two dominant gases – N_2 and O_2 – have two similar atoms in their molecules, therefore bearing no net charge in the distribution of their electrical charges during their vibration. Argon, on the other hand, is a monoatomic gas assuming no vibrational mode. These non-greenhouse gases (non-GHGs), thus, go on unaffected by infrared radiations. There are yet some other gases that have two atoms of different elements (e.g, CO and HCl) that absorb infrared radiation. However, these gases are short-lived because of their reactivity or solubility and, hence, do not contribute to the greenhouse effect to a considerable extent.

A considerable proportion of the solar radiation entering through a gaseous mantle of the Earth reaches Earth's surface. A proportion of the radiation is absorbed by the Earth's surface and the rest radiates back. The radiation reaches the Earth's surface in the form of short waves, whereas it radiates back in the form of long waves (infrared). A proportion of the short-wave radiation, while returning, is absorbed by the greenhouse gases and is used for warming up the Earth's atmosphere. The atmosphere radiates part of the energy absorbed by GHGs back to the Earth's surface (Fig. 7.1). This downward flux radiation – or, the greenhouse flux – keeps the Earth warm. This phenomenon of warming up the Earth is known as the greenhouse effect. The gases that have atoms of different elements and have a significant impact on global warming have been described in Table 7.1.

This warming up of the Earth, thanks to the role of GHGs, is necessary for the sustenance of life in the biosphere. The average temperature of the globe is 15 °C due to this greenhouse effect. Had there been no GHGs in the atmosphere, the planet's average temperature would be –20 °C, at which life in the form as exists today would almost be impossible.

Since the greenhouse effect, or warming up of the Earth's atmosphere and surface, is subject to the presence of GHGs, their atmospheric concentration relative to the non-GHGs would determine the extent of warming up of the planet – or, as it is popularly known as, the global warming.

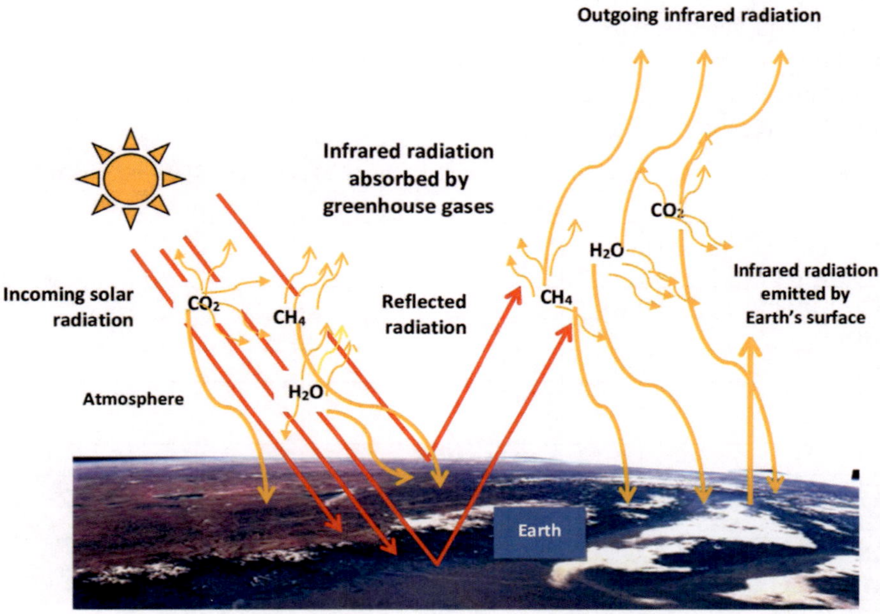

Fig. 7.1: A schematic representation of global warming by greenhouse gases

Global warming generally depends on the rise in the atmospheric concentration of the major greenhouse gas CO_2. The CO_2 functions as a glass pane of a greenhouse permitting the solar radiation to filter through it but preventing the heat from radiating back into space. A constant amount of CO_2 maintains the energy balance of solar radiation and the Earth's surface temperature. However, an increase in the CO_2 amount brings alteration in energy balance, thus contributing to a corresponding increase in Earth's surface temperature. While CO_2 plays a dominant role in global warming, other greenhouse gases also contribute to this phenomenon.

Causes of the Global Warming

Since the beginning of the Industrial Age in 1750, human activities have been increasingly adding GHGs, especially carbon dioxide, into the atmosphere. The mechanisms operated and induced by human activities adding to CO_2 and other GHGs include the overwhelming burning of fossil fuels, industrial processes, and land use changes like deforestation, agriculture, urbanization, degradation of soils, etc. As a result, there has been an increase of 48% in CO_2 from 1750 to 2019. CH_4, N_2O, and CFCs have registered an increase of 157, 20, and 262% in the same period, respectively (Table 7.1).

To judge the severity of the effects, the different GHGs have been standardized to CO_2 equivalents. These values have been determined based on the global warming potential (GWP) of a GHG over a certain period, considering a CO_2 value equal to 1 unit. The 100-year GWP for different GHGs as reported by IPCC is presented in Table 7.2.

Table 7.2: Global warming potential of some greenhouse gases over a 100-year period

GHG	Formula	100-year GWP
Carbon dioxide	CO_2	1
Methane	CH_4	28
Fossil methane*	CH_4	30
Nitrous oxide	N_2O	265
CFC-11	CCl_3F	4660
CFC-12	CCl_2F_2	10200
CFC-13	$CClF_3$	13900
CFC-113	CCl_2FCClF_2	5820
CFC-114	$CClF_2CClF_2$	8590
CFC-115	$CClF_2CF_3$	7670
HCFC-21	$CHCl_2F$	148
HCFC-22	$CHClF_2$	1760
HCFC-122	$CHCl_2CF_2Cl$	59
HFC-23	CHF_3	12400
HFC-32	CH_2F_2	677
HFC-41	CH_3F	116
HFC-125	CHF_2CF_3	3170
PFC-14	CF_4	7,390
PFC-116	C_2F_6	12,200
PFC-218	C_3F_8	8,830
Nitrogen trifluoride	NF_3	16,100
Sulfur hexafluoride	SF_6	23,500

* Metric values of methane of fossil origin include the oxidation to CO_2

Source: Compiled from the Fifth Assessment Report of IPCC (Myhre et al. 2013)

Total GHG emissions due to anthropogenic activities produced are 49 (±4.5) Gt CO_2 equivalent per year as of 2010. The GHG emission shares in six economic sectors (in percent of total human activity-related emissions) are as shown in Fig. 7.2(a). The electricity and heat production emissions shown in Fig. 7.2(b) are attributable to sectors of final energy use, or indirect emissions of electricity and heat production. Thus, adding these indirect emissions, carbon emissions from industry and buildings increase to 32 and 18.4%, respectively.

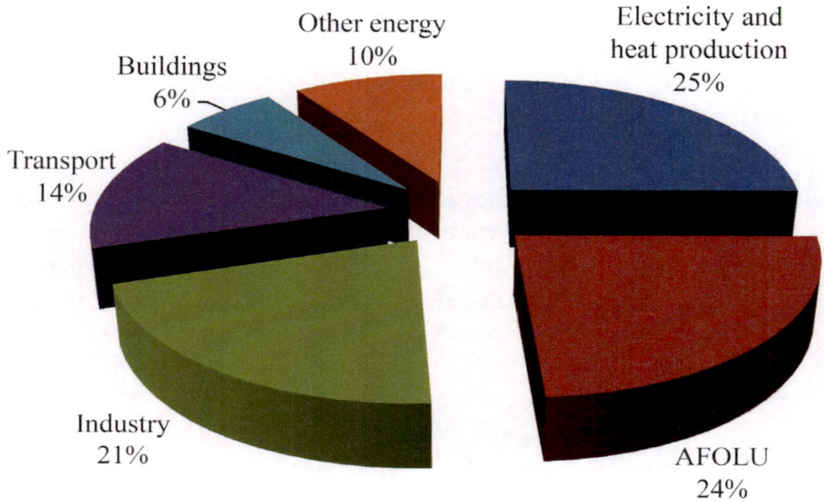

Fig. 7.2 (a): Greenhouse gas emissions by economic sectors due to human activities
AFOLU = Agriculture, forestry, and other land use

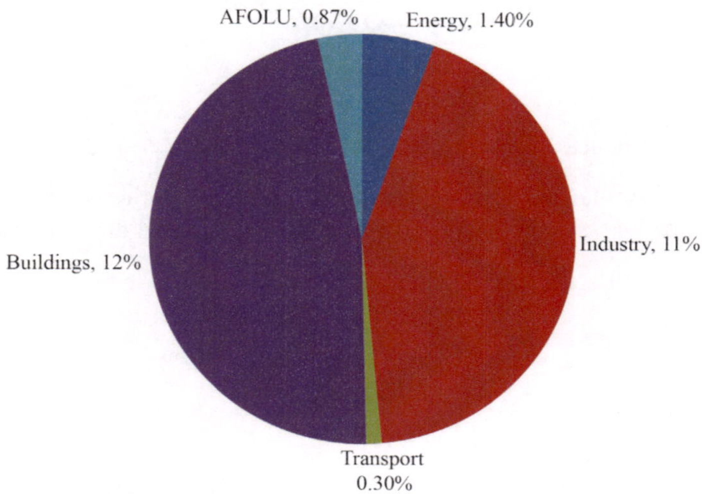

Fig. 7.2 (b): Electricity and heat production emissions attributable to sectors of final energy use

As we have observed in Table 7.2, all the greenhouse gases vary in their GWP value, which means that their impact on the process of global warming would be different. For example, N_2O has a much higher GWP value than CO_2. But, since the amount of N_2O in the atmosphere is much less than that of CO_2, its overall impact on the atmosphere would be comparatively less. The impact of each GHG is expressed in terms of the CO_2 amounts that would exert the same degree of warming.

According to IPCC (2014), of the 49 (±4.5) $GtCO_2$ eq/ yr in total GHG production thanks to human activities in 2010, the amounts of CO_2, methane, nitrous oxide, and fluorinated gases were 38 (±3.8), 7.8 (±1.6), 3.1 (±1.9) and 1.0 (±0.2) $GtCO_2$eq/yr, respectively. Thus, CO_2 remains the major anthropogenic GHG released into the atmosphere accounting for 76%, followed by methane (16%), nitrous oxide (6%), and fluorinated gases (2%) (Fig. 7.3).

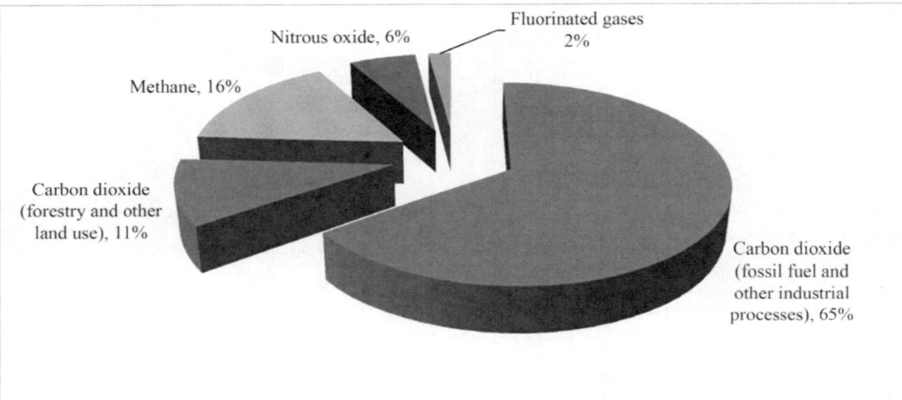

Fig. 7.3: Total GHG emissions due to human activities as of 2010 (adapted from IPCC 2014)

The state of our environment is further worsened – leading to the enhanced greenhouse effect – by various other dimensions of environmental disruptions, notably:

- Land use change
- Biodiversity erosion
- Nitrogen and phosphorus cycles
- Global freshwater use
- Ocean acidification
- Chemical changes
- Atmospheric aerosol loading
- Stratospheric ozone layer depletion

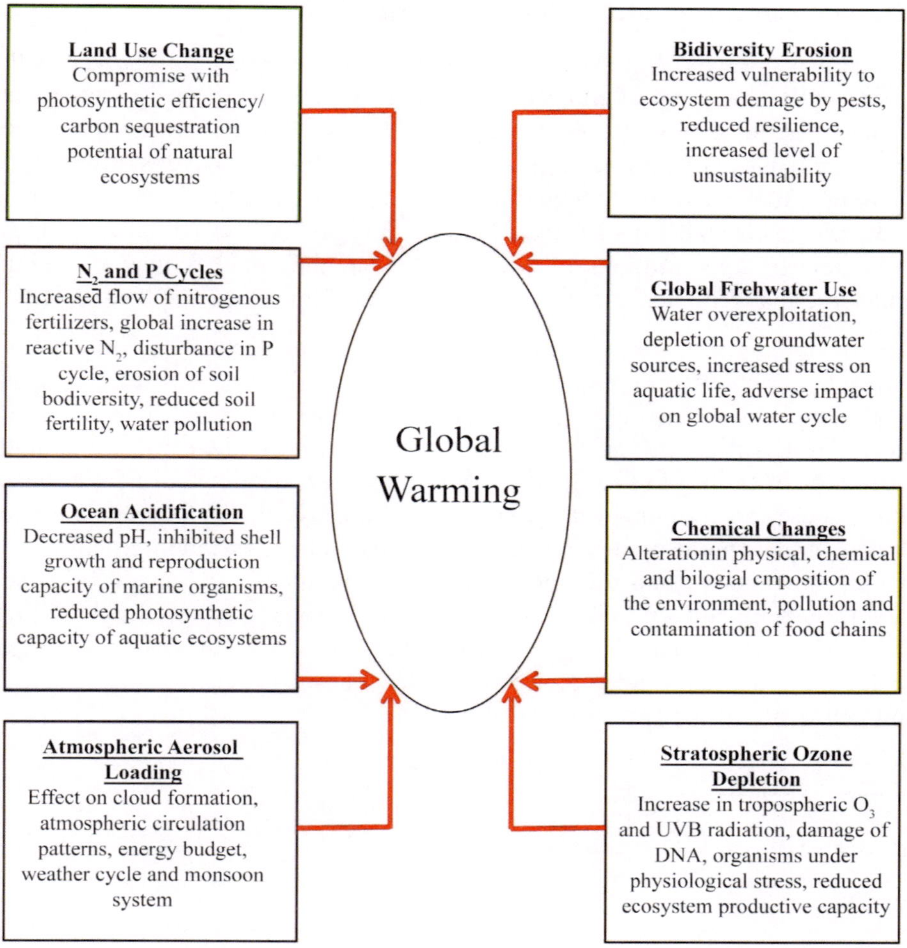

Fig. 7.4: Various dimensions of environmental disruptions exacerbating the processes of global warming and consequent climate change (adapted from Singh 2020)

These dimensions in terrestrial systems – all interrelated and having triggering effects on each other – exert severe disruptive effects on our environment, aggravating the processes of global warming and climate change (Fig. 7.4).

These changes are taking place as a result of human interventions and every change leads – directly or indirectly – to a substantial decline in ecosystems' photosynthetic capacities.

Global Warming and Climate Change

Global warming and climate change are different phenomena. However, both are closely related and often used interchangeably. While global warming, as we have already discussed, reveals the increasing global temperatures, climate change includes other more complex and specific types of changes. It involves changes in the patterns of temperature, precipitation, humidity, winds, weather, and seasonal changes over a long period. There have been climate changes in all epochs over millions of years. However, the changes that are being registered thanks to anthropogenic activities in our times are unprecedented.

We are living in an era of climate change. Climate change is principally driven by global warming. Global warming is not to halt at a point. It will, as it is doing, induce thermodynamic changes and, therefore, all dimensions of biosphere functioning. The changes that are occurring in the composition and chemistry of the Earth's atmosphere and the causes leading to induce those changes are largely human-driven as already discussed. The average climate of the planet is made up of a diversity of regional climates and microclimates. All the organisms of all the five living kingdoms of life – Monera, Protista, Fungi, Plantae, and Animalia – thrive in various climate types (Singh 2020). The variety of climates is an essential phenomenon of the biosphere that ensures the maintenance of diversity of life in terms of ecosystems, species, and genotypes of the species.

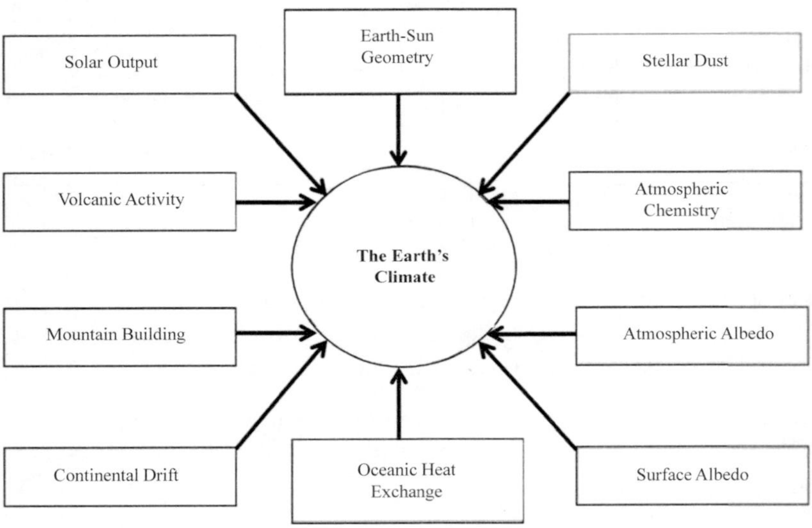

Fig. 7.5: Factors influencing the planet's climate

The anthropogenic factors we have discussed so far although dominant, need not be exclusive to contribute to climate change. Climate change is an outcome of the interactive actions of several factors. What is affected and what, in turn, affects all other interactive factors is the amount of light received on Earth's surface.

The amount of light vis-à-vis the factors that influence and are eventually influenced by light determine the Earth's climate system. There are some ten factors – cosmic as well as terrestrial – that phenomenally influence the planet's climate order as shown in Fig. 7.5.

Effects of Climate Change

Climate change is affecting and is bound to affect, the entire globe. The degree of its impact, however, may vary from region to region and area to area in the same region. The physical environment, living organisms, varieties of species, natural and anthropogenic ecosystems, and human societies are all in the range of climate change impacts.

The impacts of climate change, many of them being formidable, are already visible. The most striking ones are extreme weather events, a decline in Arctic Sea ice, glacier retreat, and sea level rise. All ecosystems of the world are registering some degree of climate change impact. Socioeconomic systems and all walks of human life are also becoming increasingly vulnerable to the undergoing climate change processes. Let us have a close look at various areas/ sectors recording climate change impacts.

Effects on Agriculture

Agriculture – including crop husbandry, forestry, and livestock production – is one of the major socioeconomic systems vulnerable to soaring atmospheric CO_2 concentrations and revealing changes in climate patterns.

Climate change is rooted in increasing atmospheric amounts of CO_2, and CO_2 is 'food' for plants. An increase in CO_2 level leads to an increase in the photosynthetic rates of green plants which is known as the CO_2 fertilization effect.

Increased rates of photosynthesis mean increased biomass production, increased water use efficiency, increased tolerance to low light levels, and increased optimum temperature for photosynthesis (Fig. 7.6).

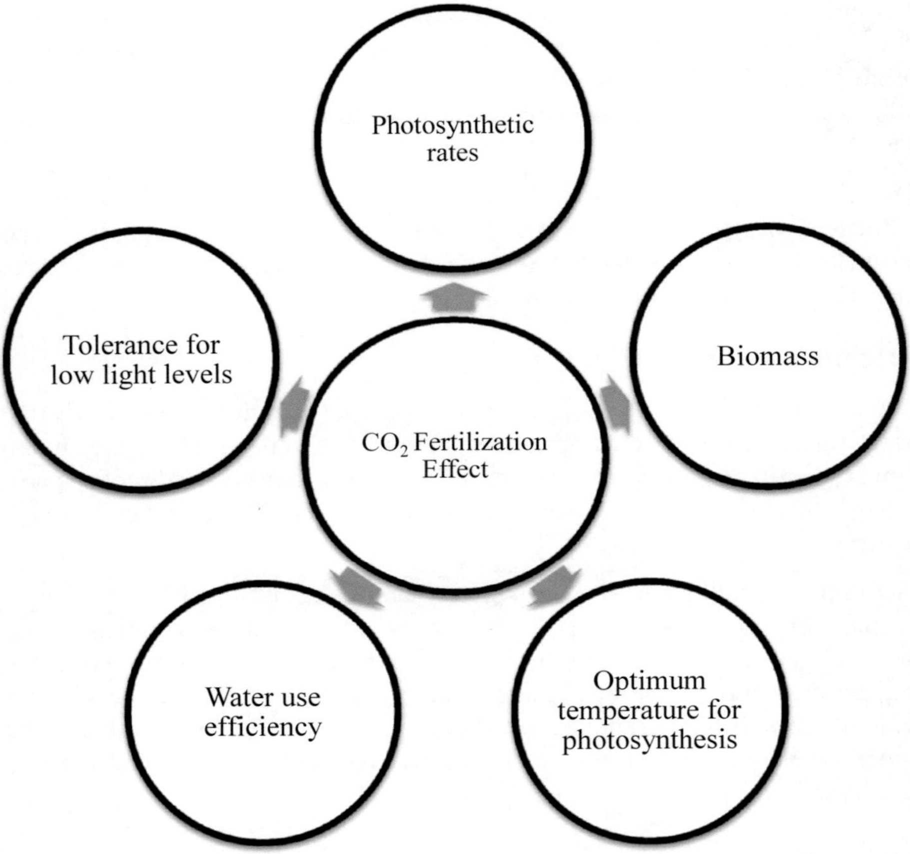

Fig. 7.6: Carbon dioxide fertilization effect on various plant parameters

The CO_2 fertilization effect depends on the availability of optimum amounts of water, minerals, temperature, and plant varieties. If all factors are available, increased CO_2 levels will be more pronounced on the C_3 plants, like wheat, rice, barley, cotton, potato, pigeon pea, etc. Since most human foods are produced from C_3 plants, it should be more likely that increasing atmospheric CO_2 would phenomenally benefit agriculture. However, the CO_2-crop production relationship is not a linear one. The CO_2 fertilization effect is likely to be hampered by the negative environmental effects of increased CO_2 amounts. The CO_2 concentrations above 400 ppm induce closure of stomata, reducing transpiration rates and hindering gaseous exchange between the atmosphere and plant leaves. This state of the environment coupled with increased temperature thanks to global warming would reduce photosynthetic rates and adversely affect plant growth and metabolic activities.

Some interesting results of CO_2-induced changes in plants' growth patterns that have so far come to the fore are: an initial but unsustainable increase in net primary productivity of forests, increased root production leading to carbon enrichment of the soil, change in plant community structure, almost no change in the microbial community, no change in leaf area index, a sharp decline in N_2 levels in forest soils.

When plants derive advantage of an initial increase in CO_2 levels up to a certain extent, there would be higher biomass production attributable to higher rates of carbohydrate synthesis, but the same would curtail the rates of protein synthesis by the plants. This state of plant growth patterns with higher CO_2 supplies would provide poor-quality foods. Also, the insect pests, to fulfill their protein needs, would extract more biomass from the plants. With such an emerging scenario, damage to crop plants by insect onslaughts would increase dramatically.

Enduring increases in CO_2 levels, in essence, would be perilous for the overall performance of plants because:

- An erratic weather cycle with too much rain over a short annual period followed by devastating floods and recurring droughts would be highly damaging for agriculture.
- Increased evaporation rates due to elevated temperatures would put the plants under water stress affecting the physiology and production performance of food crops.
- A reduced number of snowfall days would be adverse for horticultural crops in mountain areas.
- Decomposition rates of organic matter and fertilizers in the soil will increase.
- Deterioration of soil health would hamper plant growth and agronomic yields.
- The proliferation of insects and all sorts of pests would be detrimental to crop production.
- Invasion of non-native plants more adaptable to the emerging environmental conditions would intensify, usurping the original habitats of native plants/ food crops.
- Increased frequency of cyclones, hurricanes, and weather extremes would multiply adversities for agriculture.

Farming communities all over the world, especially in less developed countries, rear livestock. Domestic animals may not make a significant contribution to the global economy, but they are critical for the livelihoods of millions of farmers. Livestock contribute significantly to global warming and are themselves a victim of climate change. Animal husbandry suffers more from climate change than crop husbandry in various ways, such as (Singh et al. 2017):

1. Land use change (changes in vegetation composition and animal diets, inability of smallholders to cope with fodder deficit)
2. Reduced primary productivity of rangelands, grasslands, forages, and crops (non-availability of fodder for a longer duration, reduced nutritive value of fodder, etc.)
3. Changes in plant species composition (higher proportion/ density of invasive species in forests, rangelands, and pastures with no fodder value)
4. Quality of plant material (increased lignifications of the plants due to higher temperatures resulting in reduced digestibility of fodder)
5. The scarcity of drinking water badly affects animal health and production potential
6. Spurt in vector-borne diseases (malaria, tick-borne diseases, etc.)
7. Increased chances of helminths infection
8. Increased incidence of disease outbreaks
9. Increase in heat-related mortality and morbidity

Climate Change and Human Diseases

The ongoing climate change is intensifying human diseases and is all set to invite some unprecedented ones. As the globe warms, mosquitoes would migrate to the habitats beyond which they were currently flourishing in. In this way, they would shift the burden of several diseases such as malaria, encephalitis, meningitis, dengue fever, chikungunya, and West Nile virus (WNV) to other habitats. The scenarios of the expanding mosquito habitats would depend on the extent of climate change. So far malaria was not a threat to the people living in temperate regions and highland areas in tropical countries. With the globe becoming warmer, malaria would become a threat to those new areas. The other diseases caused by a variety of mosquitoes would also spread to new target areas.

Many of the most common infectious diseases are highly sensitive to climate variation. These include the ones transmitted by insects, vector-borne communicable diseases (e.g., malaria, dengue, hantavirus, cholera), infectious diseases (e.g., salmonellosis), giardiasis, cholera, etc. Increased temperatures and floods might lead to formidable outbreaks of these diseases. Vector-borne and communicable diseases would invade new areas as climate change goes on becoming more pronounced.

There are three disease components, viz. pathogen, host, and transmission environment (Wu et al. 2016). Humans are the important and active factors and various factors exacerbating climate change make the transmission environment favorable for the pathogens. The relationship between human infectious diseases (vector-, water-, food-, and air-borne), humans, and climate change (temperature, humidity, precipitation, wind, and pollutants) is illustrated in Fig. 7.7. From individuals, the diseases percolate into society. There is only one route then left for society to manage the diseases—adaptation.

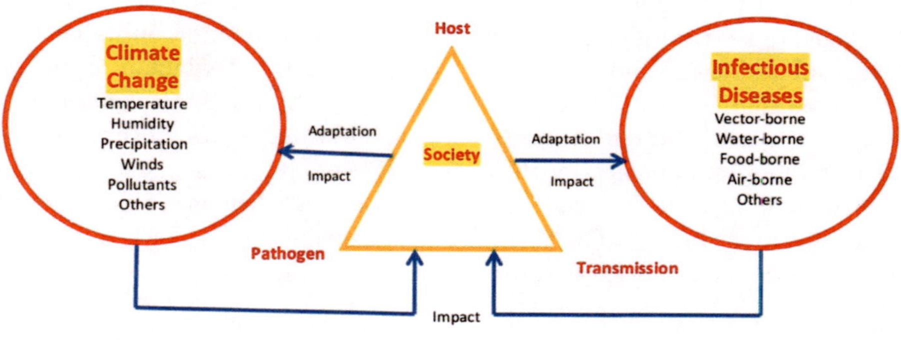

Fig. 7.7: Climate-diseases-society interrelationships

The Sixth Mass Extinction

"Extinction of species is a Law of Evolution," as Singh (2019) mentions in his theory of astrobiological evolution in his book *Fertilizing the Universe*. It has occurred in all ages ever since the planet evolved into a Living Planet. Of all the species that have ever prevailed on Earth, 99.9% have gone extinct. This is a 'culture' of the evolution. As every individual of a species dies after completing its life cycle due to apoptosis, every species also vanishes at some point in time due to the Law of Evolution. While the life span of an individual of a species is well known, the same of a species is difficult to determine. How many species have lived on Earth since the biosphere came into the folds of Biological Evolution on Earth is not known? It is vaguely estimated that only

0.1% of all the species that have come into existence in all epochs of life are living on Earth in our times.

There is a background rate or normal rate of extinction that refers to the standard rate of extinction in Earth's geological and biological history with no human interference in the Earth's ecosystems. Human species in its evolutionary processes has emerged as the most powerful one, capable to transform Earth's natural ecosystems and intrude in the natural habitats of all organisms and influence their survival. There is less likelihood that the extinction rates are still tuned to the background rates. Thus, the background rates are counted in the context of pre-human times or during periods in between major extinction events.

Species are done to extinction because of their old age—the age of a species, not of its individuals. A species generally is believed to be in existence for 1 to 2 million years for mammals and 11 million years for marine invertebrates. Apart from this generalization of the evolutionary fact, the Earth's biological organisms have experienced five mass extinctions over the geological and biological history of the Earth. The story of mass extinction on living Earth began about 440 million years ago. The Third Mass Extinction event that devoured some 96% of life on Earth about 250 million years ago is said to be the worst among these five events. The Fifth Mass Extinction event occurred some 65 million years ago in which some 75% of all the existing species were wiped out, dinosaurs included. Some experts believe that mass extinction took place due to the hitting of the Earth by a huge meteorite. A brief account of the five mass extinction events is presented in Table 7.3.

Table 7.3: The five mass extinction events spanning all geological times on planet Earth

Mass extinction	Geological time scale	Size of Extinction/ Examples of the groups affected	Possible/ Suspected cause
The First Mass Extinction The Ordovician Mass Extinction	The Ordovician Period of the Paleozoic Era about 440 million years ago	Up to 85% of all living species went extinct Examples: Brachiopods, trilobites, graptolites and moss animals	Continental drift, massive glaciations and drastic climate change that followed—the first wave during ice age and second after the end of ice age
The Second Mass Extinction The Devonian Mass Extinction	The Devonian Period of the Paleozoic Era about 375 million years ago	About 80% of all the living species vanished Examples: Coral-sponge reefs in tropics, fish and planktons	Lack of oxygen in oceans, drastic cooling of air temperatures, volcanic eruptions

The Third Mass Extinction The Permian Mass Extinction	The Permian Period of the Paleozoic Era about 250 million years ago	As many as 96% of all the species got extinct Examples: Marine invertebrates, land plants, plankton, insects along with most of life	Largely unknown, may be due to meteorite strike, volcanic eruption or unstable climate
The Fourth Mass Extinction The Triassic-Jurassic Mass Extinction	The Triassic Period of the Mesozoic Era nearly 200 million years ago	More than 50% of all the species were eliminated Examples: Large amphibians, crurotarsans, insects and conodonts	Major volcanic activity followed by basalt flooding, adverse seawater pH, sea level rise, climate change
The Fifth Mass Extension The Cretaceous Paleogene (K-T) Mass Extension	The end of the Cretaceous Period of the Mesozoic Era about 65 million years ago	About 75% of all the living species were terminated Examples: Dinosaurs (non-avian), pterosaurs, plesiosaurs, mosasaurs and ammonoids	Asteroid impact on Chicxulub Crater

The going-on sixth mass extinction is occurring at the hands of human activities. Gradually intensifying processes of adverse climate change exacerbated by human activities would catalyze the catastrophe. Occurring in the Holocene or Anthropocene epoch, this extinction event is most likely to be the worst in the history of the Living Planet. With numerous species of plants and animals in its fold, the Anthropocene extinction is estimated to occur at a rate 100 to 1000 times faster than the natural or background extinction. With its technical advancement, the human race has modified most of the Earth's ecosystems, putting all living species under control. The human species has proven itself as the global super predator engulfing anything it likes.

Humans have drastically affected all natural food chains and selectively exterminated all the species they thought might pose problems or be useless for them. In the process of overexploitation of natural resources and establishing control over all ecosystems on Earth, numerous living species today have been pushed to the brink of extinction. All the five major drivers of the sixth mass extinction illustrated in Fig. 7.8 are human-induced. The most baffling aspect of the current Holocene epoch is that just a single species – the *Homo sapiens* – holds the fate of survival of all other millions of living species. Isn't it an enigma of evolution? The human threat is not just for all non-human species but also for the human race itself.

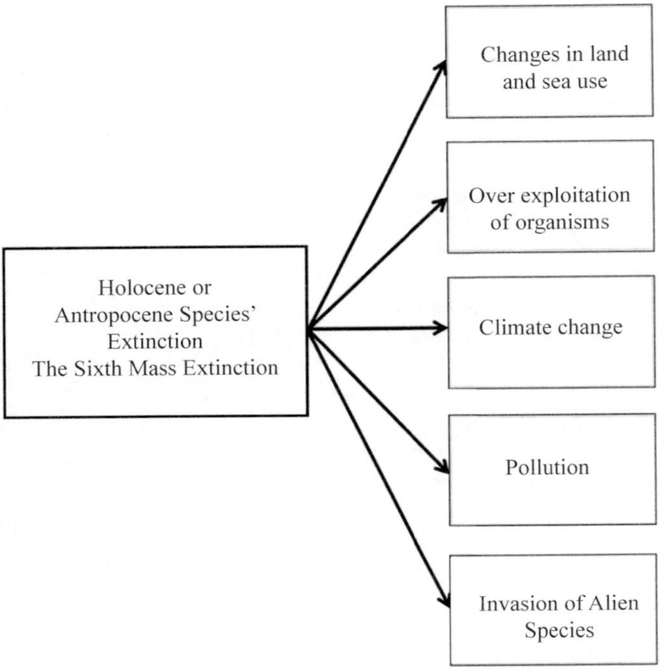

Fig. 7.8: The major drivers of the sixth mass extinction event

The Global Assessment Report of 2019 on *Biodiversity and Ecosystem Services* estimates that about one million plant and animal species face extinction within decades due to anthropogenic reasons (IPBES 2019). Earlier, Wilson (2002), based on his calculations, predicted that one-half of the Earth's higher life forms would go extinct by 2021 if the current rates of biosphere disruption by the human race continue unabatedly.

Long-Term Implications of Climate Change

The ongoing climate change is going to have very serious long-term implications. Reduced CO_2 absorption capacities of terrestrial and aquatic ecosystems would constantly contribute to accelerating CO_2 accumulation in the atmosphere. The Climate system cannot be stabilized unless there is huge curtailment in CO_2 emissions. By the time the fossil fuel-based energy resources are completely replaced by clean energy sources in compliance with global climate accords, the chances of which seem to be remote in near future, we would have been too late to halt temperature breaching the target of 1.5 to 2 °C by 2100. The symptoms arising out of soaring temperatures coupled with enhanced global warming effects will be pronounced clearly and more drastically. Negative implications of climate change on natural ecosystems (terrestrial as well as aquatic), agriculture, forestry, livestock, and all other socioeconomic sectors would be formidable.

A significant decrease in snow and ice cover on poles and outside the poles and an increase in permafrost thawing would be more pronounced indicators of climate change on the globe. Increasingly unbearable heat waves, heavy rains in summer, prolonged drought periods in arid areas, reduced flow of water in rivers and streams during drought periods would attain formidable proportions. Excessive rates of snow melt in the glaciers and poles would add to the water volumes of oceans and seas resulting in sea level rise. This state would have many human habitats in the coastal areas, including many of the world's big cities, inundated by seawater. Widespread displacement of human populations – or climate refugees – to distant places would become inevitable, adding misery to human life in many regions of the world.

The incidence of wildfires would increase dreadfully. The remaining natural ecosystems would be under severe physiological stress and primary productivity would be reduced dramatically. Agricultural productivity would fall and cropping patterns would need to be changed. In parts of Europe, Siberia, and the West, more of the land would become available for agriculture but not to derive long-term advantages.

The entire globe would be harshly affected by the gradually intensifying climate change. However, some regions would be affected more drastically than others. Earth's poles and arctic region, the Himalayas, and the Alps would be chronically affected by drastic changes in global climate patterns.

Most of the hopes to exert a cooling effect on the planet are frozen in Antarctica, Arctic, and Greenland ice sheets. However, these environments of the cryosphere are extremely vulnerable to global warming and climate change. The West Antarctic ice sheet (WAIS) may collapse within a millennium (Lenton et al. 2008) which would be one of the worst consequences of climate change creating havoc on Earth. The other of crucial environmental interest is the Arctic ice sheet about which a study by Winkelmann et al. (2015) reveals that, with cumulative fossil fuel emissions equivalent to 10,000 Gt of CO_2, this sheet could assume its complete melt over the following millennia, contributing 58 m to sea level rise, of which as much as 30 m rises would be realized within the first 1000 years. The Greenland ice sheet comprises as much freshwater in ice form as would be enough to raise sea level worldwide by 7 m, according to IPCC (2001). Greenland would be warmed enough by 2100 and its ice melt may go complete in more than 1000 years (Gregory et al. 2004, Hansen 2007).

One of the most likely dimensions of global warming to emerge would be in the form of ocean anoxia. The oxygen-dissolving capacity of the water bodies would decline considerably because of the warming of water. The higher the temperature of water the less oxygen dissolved in it. This state of the oceans would be detrimental to aquatic life.

Huge amounts of methane clathrate or methane hydrate (a form of water ice with large amounts of methane within its crystals) are lying under sediments on the ocean floor. Reserves of the methane clathrate also occur in the permafrost regions. What would happen if, in response to enduring global warming, the GHG contained in methane clathrate is released into the atmosphere? According to Archer and Buffett (2005), upon approximately 2000 Gt anthropogenic release into the atmosphere, 2000 to 4000 Gt carbon might exit from the methane clathrate reserves in the oceans and permafrost regions.

In Latin America, tropical forests would be gradually replaced by savannahs in eastern Amazonia. North America would experience decreased snow cover in western mountains. Europe would be compelled to tolerate increased risks of inland floods and high frequency of coastal floods and sea level rise coupled with increased frequency of storms would spell out devastating effects. In Asia, the Himalayan mountain range comprises the largest mass of freshwater in the form of snow and ice outside the polar regions of the Earth. As a result of long-term climate change, the lower Himalayan areas would no longer witness snowfall and the high Himalayas would have only a thin snowpack. Freshwater availability would be drastically reduced in Central and Southeast Asia.

In the Holocene epoch, the long-term climate change consequences would eventually accelerate the rates of the ongoing sixth mass extinction.

Summary

Global warming is a consequence of greenhouse gas (GHG) accumulation disproportionately in the atmosphere and is being increasingly realized and is on its way to attaining formidable magnitude. A GHG has the characteristic of absorbing and emitting radiation within the thermal infrared range. The primary GHGs in the Earth's atmosphere are carbon dioxide (CO_2), methane (CH_4), nitrous oxide (N_2O), ozone (O_3), and water vapor (H_2O). The secondary GHGs include chlorofluorocarbons (CFCs) and hydrofluorocarbons (HFCs). Although water vapor is a potential GHG contributing to the direct greenhouse effect to the tune of 36 to 72%, it is generally not a matter of concern when we discuss the fate of other GHGs.

The mechanisms operated and induced by human activities adding to CO_2 and other GHGs include the overwhelming burning of fossil fuels, industrial processes, and land use changes like deforestation, agriculture, urbanization, degradation of soils, etc. As a result, there has been an increase of 48% in CO_2 from 1750 to 2019. CH_4, N_2O, and CFCs have registered an increase of 157, 20, and 262% in the same period, respectively. The state of our environment is further worsened – leading to the enhanced greenhouse effect – by various other dimensions of environmental disruptions, notably land use change, biodiversity erosion, nitrogen and phosphorus cycles, global freshwater use, ocean acidification, chemical changes, atmospheric aerosol loading, and stratospheric ozone layer depletion.

Global warming and climate change are closely related and often used interchangeably. While global warming is an increase in global temperatures, climate change includes other more complex and specific types of changes, for example, changes in the patterns of temperature, precipitation, humidity, weather, winds, and seasonal changes over a long period. The changes that are being registered thanks to anthropogenic activities in our times are unprecedented.

The anthropogenic factors although dominant, but not exclusive to contributing to climate change. Climate change is an outcome of the interactive actions of several factors. There are some ten factors – cosmic as well as terrestrial – that phenomenally influence the planet's climate order: earth-sun geometry, solar output, stellar dust, volcanic activity, atmospheric chemistry, atmospheric albedo, mountain building, surface albedo, continental drift, and oceanic heat exchange.

The most striking impacts of climate change are extreme weather events, a decline in Arctic Sea ice, glacier retreat, and sea level rise. Socioeconomic systems and all walks of human life are also becoming increasingly vulnerable to climate change impacts. Some interesting results of CO_2-induced changes in plants' growth patterns that have so far come to the fore are: an initial but unsustainable increase in net primary productivity of forests, increased root production leading to carbon enrichment of the soil, change in plant community structure, almost no change in the microbial community, no change in leaf area index, a sharp decline in N_2 levels in forest soils, etc. The ongoing climate change is intensifying human diseases and is all set to invite some unprecedented ones. Many of the most common infectious diseases are highly sensitive to climate variation. These include the ones transmitted by insects, vector-borne communicable diseases (e.g., malaria, dengue, hantavirus, cholera, etc.), infectious diseases (e.g., salmonellosis), giardiasis, cholera, etc. Increased temperatures and floods might lead to formidable outbreaks of these diseases. Vector-borne and communicable diseases would invade new areas as climate change goes on becoming more pronounced.

The going on sixth mass extinction is occurring and being accelerated thanks to human activities eventually leading to climate crises. With numerous species of plants and animals in its fold, the Anthropocene extinction is estimated to occur at a rate 100 to 1000 times faster than the natural or background extinction. The most baffling aspect of the current Holocene epoch is that just a single species – the *Homo sapiens* – holds the fate of survival of millions of living species.

Negative implications of climate change on natural resources (terrestrial as well as aquatic), such as agriculture, forestry, livestock, and all other socioeconomic sectors would be formidable in near future. A significant decrease in snow and ice cover on poles and outside the poles and an increase in permafrost thawing, increasingly unbearable heat waves, heavy rains in summer and prolonged drought periods in arid areas, reduced flow of water in rivers and streams during drought periods would attain formidable proportions, excessive rates of snow melt in the glaciers and poles adding to the water volumes of oceans and seas resulting in sea level rise, etc. would be the consequences of climate change. The entire globe would be phenomenally affected by the gradually intensifying climate change. However, some regions would be affected more drastically than others.

Exercises

A. Multiple Choice Questions

1. Which of the following groups does represent non-greenhouse gases?
   a. N_2O, CH_4, O_3, CFCs
   b. CO_2, CH_4, HFCs, H_2O vapor
   c. CO_2, CH_4, N_2O, O_3, H_2O vapor, CFCs, HFCs
   d. N_2, O_2, Ar
2. What was the atmospheric CO_2 concentration at the beginning of the Industrial Age, 1750?
   a. 180 ppb
   b. 280 ppb
   c. 280 ppm
   d. 380 ppm
3. What was the concentration of CFCs in the pre-Industrial Age?
   a. 262 ppt
   b. 262 ppb
   c. 262 ppm
   d. Zero

4. Nitrous oxide will be released mainly when
   a. Nitrogenous fertilizers are decomposed in the soil
   b. Surfur-rich organic matter is burnt
   c. Anaerobic decomposition of organic matter takes place
   d. Methanogens decompose organic matter in the garbage
5. Which of the following gases is mainly released from enteric fermentation?
   a. O_3
   b. N_2O
   c. CFCs
   d. CH_4
6. Which of the following is a greenhouse gas?
   a. Sulfur hexafluoride
   b. Tetrafluoromethane
   c. Hexafluoroethane
   d. All of the above
7. The greenhouse effect is due to
   a. Permeability of short wavelength radiation through the ozone layer
   b. Impermeability of long wavelength radiation through GHGs
   c. Permeability of short wavelength radiation through GHGs
   d. Impermeability of long wavelength radiation through the ozone layer
8. On 100 yearly basis, which of the following greenhouse gases will carry the maximum GWP?
   a. Methane (CH_4)
   b. Fossil methane (CH_4)
   c. CFC-11 (CCl_3F)
   d. Nitrous oxide (N_2O)
9. Marginal increase in one of the following greenhouse gases is likely to have a positive bearing on the performance of food crops via enhancement of photosynthetic rates.
   a. CO_2
   b. CH_4
   c. N_2O
   d. CFC-11
10. Which of the following human activities would contribute to global warming?
    a. Rice cultivation
    b. Livestock farming
    c. Deforestation
    d. All of the above

11. The West Nile, in the context of the consequences of global warming on human health, is referred to as a kind of
    a. Bacteria
    b. Virus
    c. Crab
    d. Water pollution of the Nile River
12. The normal rate of extinction in Earth's geological and biological history with no human interference in the Earth's ecosystems refers to
    a. Background rate
    b. Species' annihilation rate
    c. Self-extinction rate
    d. Geological standard rate
13. The First Mass Extinction of the species on our planet refers to the
    a. Ordovician Mass Extinction
    b. Devonian Mass Extinction
    c. Permian Mass Extinction
    d. Triassic-Jurassic Mass Extinction
14. Considering the proportion of the species extinction, which of the mass extinctions in the Earth's biological history was the worst?
    a. The First Mass Extinction
    b. The Second Mass Extinction
    c. The Third Mass Extinction
    d. The Fourth Mass Extinction
15. Which mass extinction is held responsible for wiping out dinosaurs?
    a. The Cretaceous Paleogene (K-T) Mass Extension
    b. The Triassic-Jurassic Mass Extinction
    c. The Permian Mass Extinction
    d. The Devonian Mass Extinction
16. The Holocene or the Anthropocene species extinction is the
    a. Sixth Mass Extinction
    b. Fifth Mass Extinction
    c. Fourth Mass Extinction
    d. Third Mass Extinction
17. A form of water ice with large amounts of methane within its crystals on the ocean floor is known as
    a. Methane ice
    b. Methane clathrate
    c. Geological methane
    d. Volcanic methane

18. The Antarctic ice likely to melt completely within a millennium due to adverse climate changes is known as
    a. North Antarctic ice sheet (NAIS)
    b. West Antarctic ice sheet (WAIS)
    c. East Antarctic ice sheet (EAIS)
    d. South Antarctic ice sheet (SAIS)
19. This ice sheet of an island comprises as much freshwater in ice form as would be enough to raise worldwide sea level by 7 m, according to IPCC. This is
    a. Borneo ice sheet
    b. Greenland ice sheet
    c. Himalayan ice sheet
    d. Alps ice sheet
20. In Asia, the largest freshwater reserve in the form of snow and ice is there in the
    a. Central Asia
    b. Hindu Kush Mountains
    c. Himalayas
    d. Mount Kinabalu

Answers: 1-**d**, 2-**c**, 3-**d**, 4-**a**, 5-**d**, 6-**d**, 7-**b**, 8-**c**, 9-**a**, 10-**d**, 11-**b**, 12-**a**, 13-**a**, 14-**c**, 15-**a**, 16-**a**, 17-**b**, 18-**b**, 19-**b**, 20-**c**

B. **Short-Answer Questions**

1. What is a greenhouse gas?
2. Name the primary greenhouse gases.
3. What is the greenhouse effect?
4. Which GHG plays a major role in global warming?
5. What do you mean by the global warming potential (GWP) of a greenhouse gas?
6. Mention the human activities exacerbating the greenhouse effect.
7. What is the difference between global warming and climate change?
8. What socioeconomic sectors play role in climate change?
9. Mention the names of some human diseases that might emerge due to climate change.
10. Name the mass extinction events that have ever taken place on earth.

C. Long-Answer Questions

1. Describe through illustration how global warming is caused.
2. Describe the effects of climate change on agriculture and livestock.
3. What are the possible terrestrial and non-terrestrial factors that may be contributing to climate change?
4. What are the long-term implications of climate change? Describe giving examples.
5. Write short notes on:
    a) Causes of Global Warming
    b) Climate Change Effect on Human Diseases
    c) The Sixth Mass Extinction

Suggested Readings

Archer, S. and Buffett, B. 2005. Time-dependent response of the global ocean clathrate reservoir to climatic and anthropogenic forcing. *Geochemistry Geophysics Geosystems*, 6 (3): Q03002.

Chen, C., Park, T., Wang, X., Piao, S., Xu, B., Chaturvedi, R. K., Fuchs, R., Brovkin, V., Ciais, P. Fensholt, R., Tommervik, H., Bala, G., Zhu, Z., Nemani, R. R. and Myneni, R. B. 2019. China and India lead in greening of the world through land-use management. *Nature Sustainability* 2: 122-129. https://doi.org/10.1038/s41893-019-0220-7

Gregory, J. M., Huybrecht, P., Raper, S. C. 2004. Climatology: threatened loss of the Greenland ice-sheet. *Nature*, 428 (6983): 616.

Hansen, J. E. 2007. Scientific reticence and sea level rise. *Environmental Research Letters* 2 (2): 024002.

IPBES (Intergovernmental Science-Policy Platform on Biodiversity and Ecosystem Services). 2019. The Global Assessment Report on Biodiversity and Ecosystem Services – Summary for Policymakers. Bonn, Germany: IPBES Secretariat. 56 pp. https://doi.org/10.5281/zenodo.3553579

IPCC (Intergovernmental Panel on Climate Change). 2001. Climate Change 2001: The Scientific Basis. Contributions of Working Group I to the Third Assessment Report of the Intergovernmental Panel on Climate Change [Houghton, J. T., Ding, Y., Griggs, D. J., Noguer, M., van der Linden, P. J., Dai, X., Maskell, K. and Johnson, C. A. (eds.)]. Cambridge and New York: Cambridge University Press. 881 pp.

IPCC (intergovernmental Panel on Climate Change). 2001. Climate Change 2001: The Scientific Basis. London and New York: Cambridge University Press.

IPCC (Intergovernmental Panel on Climate Change). 2014. Summary for Policymakers. In: Climate Change 2014: Mitigation of Climate Change. Contribution of Working Group III to the Fifth Assessment Report of the intergovernmental Panel on Climate Change

[Edenhofer, O., Pichs-Madruga, R., Sakona, Y., Farahani, E., Kadner, S., Seyboth, K., Adler, A., Baum, I., Brunner, S., Eickemeier, P., Kriemann, B., Savolainen, J., Schlomer, S., von Stechow, C., Zwickel, T. and Minx, J. C. (eds.)]. Cambridge and Newy York: Cambridge University Press.

Keenan, T., Prentice, I., Canadell, J. G., Williams, C. A., Wang, H., Raupach, M. and Collatz, G. J. 2016. Recent pause in the growth rate of atmospheric CO_2 due to enhanced terrestrial carbon uptake. *Nature Communications* 7, 13428. https://doi.org/10.1038/ncomms13428

Lenton, T. M., Held, H., Kriegler, E., Hall, J. W., Lucht, W., Rahmstorf, S. and Schellnhuber, H. J. 2008. Inaugural article: Tipping elements in the Earth's climate system. *Proceedings of National Academy of Sciences* 105 (6): 1786-1793.

Myhre, G., Shindell, D., Breon, F.-M., Collins, W., Fuglestvedt, J., Huang, J., Koch, D., Lamarque, J.-F., Lee, D., Mendoza, B., Nakajima, T., Robock, A., Stepens, G., Takemura, T. and Jhang, H. 2013. Anthropogenic and natural radiative forcing. In: Climate Change 2013: The Physical Science Basis. Contributions of Working Group I to the Fifth Assessment Report of the Intergovernmental Panel of Climate Change [Stocker, T. F., Quin, D., Plattner, G.-K., Tignor, M., Allen, S. K., Boschung, J., Nauels, A., Xia, Y., Bex, V. and Midgley, P. M. (eds.)]. Cambridge and New York: Cambridge University Press.

Norby, R. J. and Zak, Donald R. 2011. Ecological lessons from free-air CO2 enrichment (FACE) experiments. Annual Review of Ecology, Evolution, and Systematics 42 (1): 181-203.

Singh, Vir, Rastogi, A., Nautiyal, N. and Negi, V. 2017. Livestock and climate change: the key actors and sufferers of global warming. *Indian Journal of Animal Sciences* 87 (1): 11-20.

Singh, Vir. 2019. *Fertilizing the Universe: A New Chapter of Unfolding Evolution*. London, U. K.: Cambridge Scholars Publishing.263 pp.

Singh, Vir. 2020. *Environmental Plant Physiology: Botanical Strategies for a Climate Smart Planet*. Boca Raton and New York, USA and London, UK: Taylor and Francis (CRC Press). 214 pp.

Wilson, E. O. 2002. *The Future of Life*. New York: Vintage Books. 256 pp.

Winkelmann, R., Levermann, A., Ridgwell, A. and Caldeira, K. 2015. Combustion of available fossil fuel resources sufficient to eliminate the Antarctic Ice Sheet. *Science Advance* 1 (8): e1500589.

Wu, X., Lu, Y., Zhou, S., Chen, L. and Xu, B. 2016. Impact of climate change on human infectious diseases: Empirical evidence and human adaptation. *Environmental International* 86: 14-23. https://doi.org/10.1016/j.envint.2015.09.007

Web Resources

https://byjus.com/biology/global-warming/

https://coastadapt.com.au/causes-of-climate-change-and-sea-level-rise

https://earthobservatory.nasa.gov/features/GlobalWarming

https://education.nationalgeographic.org/resource/global-warming

https://news.un.org/en/story/2019/05/1037941

https://www.britannica.com/science/climate-change

https://www.discovery.com/nature/99-Percent-Of-The-Earths-Species-Are-Extinct

https://www.nasa.gov/audience/forstudents/k-4/stories/nasa-knows/what-is-climate-change-k4.html

https://www.strategian.com/2020/08/08/the-sixth-mass-extinction-on-earth/

https://www.un.org/en/climatechange/what-is-climate-change

https://www.vedantu.com/biology/global-warming

https://www.who.int/health-topics/climate-change#tab=tab_1

Glossary

Aerosols the solid or liquid particles or both with diameter mostly less than 1 μm and behaving like gases when suspended in the air.

Agrochemicals the chemical products, such as synthetic fertilizers and pesticides used in agriculture.

Becquerel (Bq) the International System (SI) of the measure of radioactive decay. One *Becquerel* is activity of the quantity of a radioactive material in which one nucleus decays.

Biological magnification the phenomenon through which certain pollutants get accumulated in organisms in increasing concentrations along the food chain in trophic hierarchy.

BOD (biochemical oxygen demand) the measure of DO needed by decomposers in the decomposition of organic matter present in water; it is expressed in mg of O_2 per liter of water.

COD (chemical oxygen demand) a measure of oxygen equivalent of the requirements for the complete oxidation of total organic substances, biodegradable as well as non-biodegradable, present in the water.

Curie (Ci) a measure of radioactivity; a Curie being the amount of radioactivity from 1 g of Radium-226 that undergoes about 37 billion nuclear transformations per second.

DDT (dichlorodiphenyltrichloroethane) a chemical compound, an organochloride, originally developed as an insecticide; it is often referred on account of its serious environmental effects.

Decibel (dB) a measure of sound intensity by a ratio of loudness with reference to the softest audible sound.

DO (dissolved oxygen) a measure of oxygen dissolved in water.

Eutrophication the process of nutrient enrichment of a water body leading to changes in species composition, density of populations and loss of biodiversity.

Global warming potential (GWP) the ability of a greenhouse gas to trap heat in the atmosphere compared to CO_2.

Global warming the long-term rise in the average temperature of the Earth's climate system.

Greenhouse effect warming of the planet due to trapping of heat by the gases known as greenhouse gases, especially CO_2, H_2O, CH_4, N_2O, O_3, CFCs and HFCs.

Greenhouse gases the atmospheric gases, such as CO_2, H_2O vapour, CH_4, N_2O, O_3 etc. that hold Earth's surface temperature within a range appropriate for the physiological wellbeing and functioning of living organisms, but that also contribute to elevate temperature if their concentration increases disproportionately.

Ionizing radiation any particle or electromagnetic wave carrying as much energy as will be enough to ionize or remove electrons from an atom; e.g., particle radiation from radioactive decay (Alpha- and Beta-particles and neutrons) and Gamma rays, X-rays, and far ultraviolet light.

Methane clathrate or methane hydrate, a form of water ice with large amounts of methane within its crystals, lying under sediments on ocean floor which may release into the atmosphere due to enduring global warming. Reserves of the methane clathrate also occur in the permafrost regions.

Non-ionizing Radiations the short-wave radiations as part of solar radiations, such as ultraviolet rays. These radiations have low penetration power and are harmful to multicellular organisms when they are exposed to such radiation for a longer period.

Ocean anoxia a condition of oxygen depletion in the oceans; the on-going global warming may induce this condition by gradually reducing water-dissolving properties of the oceans.

Pathogen an organism – commonly among viruses, bacteria and protozoa – that induces disease, a debilitating condition, in its host.

Peroxyacyl nitrates (PANs) a family of phytotoxins, with PAN as the most abundant member of the family, in the photochemical smog, causing damage to chloroplasts and consequently impairing photosynthesis, electron transport chain of respiration, enzyme activity and anabolic processes in plants.

Pollution an undesirable (or negative) change in the physical, chemical and biological characteristics of the environment or of a component of the environment (air, water, soil etc.) occurring directly and/ or indirectly that may affect life adversely in one way or the other or may carry potential to harmfully affect life instantly or over a period of time.

Radiation the transfer of heat through electromagnetic radiation.

Radioactive waste a hazardous waste containing radioactive material.

Reactive oxygen species (ROS) the $\bullet O_2^-$, $\bullet OH$ and H_2O_2 produced as a result of oxygen reduction by electron transfer system of mitochondria in aerobic organisms, together with unstable intermediates in the peroxidation of lipids.

Salinity the salt content of water; denotes the amount of dissolved salt in a water body.

Suspended particulate matter (SPM) the finely divided solid and liquid particles that disperse through atmosphere causing pollution.

The Sixth Mass Extinction an ongoing species extinction event during the present Anthropocene or Holocene epoch.

Zooplankton the animals, mostly microspcopic, that drift in the surface waters of the lakes or oceans.

Subject Index

N

O

P

U

V

W

X

Z